Consulting Engineering: constructing the future

ENGINEERING MANAGEMENT SERIES

Series Editor: **Dr. John A. Brandon**
University of Wales, Cardiff, UK

Consulting Engineering: constructing the future

Peter M. Hartley

Formerly Chief Executive of DHV (UK) Ltd.

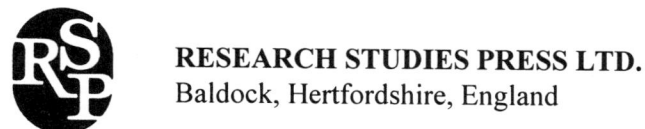

RESEARCH STUDIES PRESS LTD.
Baldock, Hertfordshire, England

RESEARCH STUDIES PRESS LTD.
15/16 Coach House Cloisters, 10 Hitchin Street, Baldock, Hertfordshire, England, SG7 6AE

and

325 Chestnut Street, Philadelphia, PA 19106, USA

Marketing:

UK, EUROPE & REST OF THE WORLD
Research Studies Press Ltd.
15/16 Coach House Cloisters, 10 Hitchin Street, Baldock, Hertfordshire, England, SG7 6AE

Distribution:

NORTH AMERICA
Taylor & Francis Inc.
International Thompson Publishing, Discovery Distribution Center, Receiving Dept., 2360 Progress Drive Hebron, Ky. 41048

ASIA-PACIFIC
Hemisphere Publication Services
Golden Wheel Building, 41 Kallang Pudding Road #04-03, Singapore

UK, EUROPE & REST OF THE WORLD
John Wiley & Sons Ltd.
Shripney Road, Bognor Regis, West Sussex, England, PO22 9SA

Library of Congress Cataloging-in-Publication Data
Hartley, Peter M., 1941-
    Consulting engineering : constructing the future / Peter M. Hartley.
        p. cm.
    Includes bibliographical references and index.
    ISBN 0-86380-209-5
        1.   Consulting engineers. 2. Engineering firms--Management. I. Title.

    TA216.H37 2000
    620'.0023--dc21

                                                                99-9057568

British Library Cataloguing in Publication Data
A catalogue record for this book is available from the British Library.

ISBN 0 86380 209 5

Printed in Great Britain by SRP Ltd., Exeter

DEDICATION

To my wife, friend and partner
Fiona

without whose constant encouragement this book
would never have been completed.

OFFERING

This book is offered to innumerable colleagues, clients
and competitors who may recognise some of their thoughts
and our discussions in the text.

I offer it to them in the hope that it will be useful
for the future of the profession of Consulting Engineering - however
difficult things may at times appear!

"All things will be well, and all manner of things will be well"
Julian of Norwich

West Hoathly,
West Sussex.
United Kingdom RH19 4QF

ACKNOWLEDGEMENTS

My thanks are due to many people who have encouraged me to write this book and to set out my thoughts about the profession of Consulting Engineering.

They have provided some of the information, thoughts and ideas which form parts of the book. I have tried to make due acknowledgement within the text where this seemed appropriate; if I have not, I trust that they will forgive the omission. I should also like to particularly acknowledge those who have responded to detailed enquiries, many of whom are listed in Table 3.6.

But most of my contact with people and companies has been on an informal basis and many have given me time to discuss my thoughts with them; I should like to thank them all for their availability, help, courtesy and patience.

CONTENTS

CHAPTER 1

Background to the book

1.1. THE IDEA OF THE BOOK

For more than 30 years I have worked as a Consulting Engineer, mainly for British firms but also with European companies, designing and managing civil engineering projects and, more recently, managing these companies. Projects that I've worked on have covered the UK but they have also taken me extensively to the Middle East, North Africa and mainland Europe. For a long time I've had that nagging urge to write about the industry, its ways of doing things, and its future and so this book comes from the heart!

It is about an industry and profession which I have loved and which has given me, personally, many satisfying and enjoyable times. The construction industry in the UK is full of many colourful characters and the consultancy side of the industry is equally full of fascinating and creative, lively people, albeit perhaps a little less flamboyant than some of their contractor counterparts! A friend of mine said recently, and tellingly, *"People in engineering consultancy have to be tough and really resilient, if they are to survive in the climate which we work in"*.

If any industry could be described as a real people industry, then this must be the one; it has a wealth of fascination in its mix of machines, projects, men, and, nowadays, women. And the pattern is repeated, from my experience, all over the world; construction has a 'pull' about it which is hard to beat.

Yet the whole industry is, I believe, feeling increasingly uncomfortable about its future because it is being asked to change in many ways and at increasing speed so as to survive in anything like its present size and meet the challenges of our new century. This is certainly true for the UK situation and for construction in many other parts of the world. Within this overall canvas, the work of Consulting Engineers is changing as much as any other element; this has been my background and my love for many years, working with a rich variety of people both in the UK and in other parts of the globe.

The profession of the Consulting Engineer has, traditionally, been one that has been regarded with respect world-wide; that respect has particularly brought to British consultants a dominance in this field through much of the past 100 years. Major firms have worked almost everywhere, undertaking projects and transferring their skills to nationals in other countries, whilst at the same time gaining for themselves a broad base of expertise that has enabled them to weather the cyclical variations in markets and economies, particularly those in their home territory.

A fairly similar pattern has applied to large firms of Consulting Engineers in most countries of mainland Europe, although their spread of work outside their home territory has tended to be smaller than for many a UK counterpart and their development in the international scene took longer to get going.

In North America, however, engineering consultancy has had very different origins and there is, altogether, a different history and often a very concept of project development. They have not, traditionally, worked very much outside their own continent as consultants in their own right, but it is noticeable that this has begun to change over the last 10-15 years; now more work is sought outside North America using a more European model of consultancy, and acquisitions and linkages have been made with European firms, often for this specific purpose.

In other parts of the world, e.g. South-east Asia, Southern Africa and Australasia, models of consultancy have tended to follow the European pattern, but with the vast majority of the work of these Consulting Engineers being carried out within their own domestic environment, certainly rarely outside their native region.

All of these patterns are now changing and the era of dominance for British consultants has now long passed and the profession is at a cross-roads, I believe, threatened on many sides, not least, some would say, in its own home market. The industry needs significant change if it is to move forward and flourish in the new millennium. Engineers, whilst not resistant to change, are not always renowned for their flexibility, innovation and creativity outside the particular technical specialism in which they excel. Resistance to change in this, and indeed in most industries, is prompted by fear of the unknown - something that touches and inhibits almost all of us.

But change it must; and, in my view, quickly!

This book is an attempt to look into that future - largely unknown - so as to identify the changes that are appropriate for the profession, both in the UK and elsewhere, which will help Consulting Engineers to be confident about their destiny and the future.

1.2. THE BACKGROUND

What is driving this need for change?

This evidence stares us straight in the face, although many a statistician would regard it as anecdotal rather than statistically robust. Yet for those in the industry, it is as real as ever any trend could be. The changes amongst British firms of Consulting Engineers over the last 10 years have been dramatic, to say the least; they are part of a dramatic shake-out within the whole of the construction industry which started in the mid-1980s and which shows every sign of going on just as much in the new millennium. Indeed, for construction as a whole, *"the industry is facing its most challenging times from new and dynamic competitors."*[1]

During this period, very few firms have been untouched by the wave of mergers, acquisitions, take-overs and disposals that have occurred; many of these have involved international companies, and now the industry, generally, has a much more complex and varied ownership pattern. The public sector, and much of private industry, both with large in-house teams of professionals, have given up their traditional and protected client bases and have started to compete right across the market, sometimes as part of, and sometimes in competition with, long-standing firms of consultants. This is all indicative of a sector in considerable turmoil.

At the same time, the market, world-wide, but particularly in Europe, has become much more open and competitive; perhaps the UK market is the most extreme of those affected in this way. Clients also have become more demanding, no longer entrusting entire projects to a single firm, but requiring a complex structure of multi-professional relationships and searching for that magic quality of 'total value for money' which is so often elusive. Innovative forms of setting up projects are on the increase, requiring an ever more intricate web of disciplines and relationships, and large up-front costs for major projects.

---

[1] Letter by **Malcolm Knobel-Forbes** in Management Today, April 1998, in response to a feature on the Construction Industry, February 1998

Yet, many of these aspects are mirrored, in one form or another, right across both the manufacturing and service sectors of industry. Everywhere there are new forms of working, new ways of managing 'people businesses', new types of ownership and new technologies, coming at us with increasing speed. Consultants, of all engineering disciplines, with their 'people power', can be in the forefront of these developments.

Within this setting, another major factor in all the uncertainty is that the total market in the UK has been static or in decline for a number of years, leading to shake-outs and the restructuring of long-established operations amongst consultants as well as a frantic search for new fields of activity and business. Fee levels have fallen significantly, increased project input is demanded, and generally all work is now done, and re-done, at a much faster pace than in years gone by.

The result of these trends has been all too plain to see, with many consultancies finding it quite a struggle to survive, let alone make sufficient profit to ensure the continual investment which is so essential for the future. Yet, at the smaller end of the scale, barely a day seems to go by without news of a new entry into this over-crowded, and sometimes apparently desperate, market. Many of these new entrants are young, dynamic and very ambitious; they embrace the very latest ways of working and technology and they can be real threats to established consultancies.

1.3. THE APPROACH

The broad approach of this book is to try to track and identify many of these trends, to look at how other industries are coping with similar, yet obviously different problems, and to see how the profession might move forward within the context of a very changing market and new technologies.

It looks at the patterns of future markets for consultants, particularly in the UK, to see how the changing market demands can be met most effectively. It examines some of the current themes in client requirements and the response that should be made to these, for, in this age as never before, the client is "king".

It goes on to look at future employment and working practices and at organisational structures that might be right for engineering consultancies, wherever they operate, and which should bring out the best in their people; for it is the skills and abilities of the people in any consultancy that are its only real and fundamental assets.

As with any book of this type and about an industry so varied as engineering consultancy, the themes developed and the illustrations given can only be a snapshot of what is happening and what has to happen throughout the industry, not just in the UK but, increasingly, worldwide. Inevitably they will be incomplete and often inappropriate for a particular company, for there is no set or single solution which will provide the new future for every organisation within the industry.

What I hope the book will do is to flag up sufficient issues in enough detail for current practitioners, those who manage and work in engineering consultancy, to take at least some of them on board. For I believe that this will then help them to look at what they are doing in the wider light of their industry and new working practices as a whole; we are all in a massive period of change and it is, most of the time, quite difficult to get the total grasp of it.

Many similar trends can be seen within consultancy in many parts of the world and, indeed, within many other professions and types of service businesses. I would hope that some of the ideas in this book might well be relevant to other businesses where there are substantial numbers of one-off projects all involving a mix of people skills and technology.

One of the problems of many management books is that they're either academic, and thus do not really appeal to practitioners, or they tend to use, as illustrations, examples from major multi-national corporations, which most Consulting Engineers are not. The intention of this book, in contrast, is to take some relevant management and industry ideas, tempered by experience and some 'looks into the future,' and then apply them in a more practical way to this particular industry.

The book therefore follows a pattern:-

- Where is engineering consultancy now and what changes have happened in the market during recent years?
- What is happening in the wider world of all service industries and employment; what innovations are being tried out and how do these work in practice?
- How are future markets for Consulting Engineers changing?
- How can Consulting Engineers adapt their working practices for the new market, for their performance objectives, and to meet the aspirations of their employees, clients and other stakeholders?

The book is not a blueprint for future success in the engineering consultancy industry, but I hope, at least, that the adoption of some of its strategic thoughts and ideas will help firms to shape a prosperous future for themselves and for the industry as a whole.

Because it has this structure, the reader may well want to note appropriate ideas at the end of each chapter; space is left for this.

NOTES ON THE CHAPTER

CHAPTER 2

History and the changing market

2.1. GENERAL BACKGROUND

2.1.1. The early days

It's generally accepted, I believe, that British Consulting Engineers were very much the founders of the present profession and thus they led the way in developing the profession from its earliest days. Their example was followed in many other countries, but with obvious and justifiable national variations. In order to understand fully the position that Consulting Engineers are now in, it's therefore appropriate to look back in some general ways at their early history. I believe that this also helps to explain some current attitudes that are found in the profession, particularly amongst consulting civil engineers in the UK, but also elsewhere.

Going back a long time, to the founding fathers in civil engineering in the period 1830-1860, men like Brunel and Stephenson were entrepreneurs, often promoting major engineering projects in their own right. At the same time, they were also employed as the Engineer for the project by, for example, the early railway companies, who had themselves raised funds from wealthy sponsors for their enterprise; this then became a role model for the Consulting Engineer that survived largely unchanged for the next 100 years.

It was in the latter part of the 19th century that several of the now household names in UK engineering consultancy were founded, mainly civil engineers, and they worked both in their home market and in the (then) British Empire. This pattern of activity continued for the following 50-75 years through the recessions of the late 19th century and the 1930s. It was also during this period that other well-known consultancies were established outside the UK, sometimes working in a similar way to those in Britain, for example in the Netherlands and Germany, but also developing a characteristic way of working of their own, for example in France and North America.

With the ending of the 2nd World War, a new generation of Consulting Engineers set up in practice, often on the back of wartime experience. Other individuals, however, in the 1950s and 1960s, were excited at the prospect of establishing their own firms, having worked first for some of the larger and long-established practices.

As the UK industry grew in this way, so definite patterns of work began to become clear. Most of the larger and long-established practices operated on a world-wide scale with several in fact having the majority of their work abroad; this provided them with a wider and more stable base to counteract the 'ups and downs' in the British economy which affected their domestic workload. It also provided, in many instances, virtually automatic continuity of work on a series of major projects which kept the UK offices fully employed perhaps for decades. The smaller and newer firms, however, tended to concentrate their efforts within the home market, although there were some notable exceptions to this.

Within the home market, most of the larger firms of consultants worked almost entirely in the public sector, which accounted for the bulk of major projects; the private sector was definitely the preserve of the smaller and generally newer firms, some of whom had a balance in their workload, with others working almost entirely for private clients. There was a similar pattern of work in most of the developed countries.

2.1.2. The first signs of change in the UK

It was in the late 1960s that the first large changes in the pattern of working for British Consulting Engineers began to occur; this was the start of a process that has been going on ever since.

Rivalry between the traditionally well-established firms had always been present, but in a fairly gentlemanly way; now the new, upstart, consultancies started to make their presence known in the market by seeking to obtain projects which hitherto had been very much the preserve of the more long-established firms. Within the public sector, where the larger firms did most of their business, there began to be encouragement for the smaller, newer firms who perhaps came across as more responsive; this took the form of granting them smaller commissions initially, which naturally progressed, in time, to more substantial contracts. At the same time, some government sector clients took a keen interest in promoting firms which were seeking to expand their export effort, these firms pointing out that a reasonable amount of home-based work was essential for their credibility if their exporting effort was to be successful.

In this way, competition across the industry, certainly in the public sector, intensified, although, as yet, there was little competition on a fee basis; that was to come later - with a vengeance. The general impression gained of this period was that public sector clients were attempting in many cases to be seen to be fair in their award of commissions; often, however, this appeared, in practice, to favour the more extrovert and entrepreneurial consultants who, in turn, tended to represent the newer players in the industry.

The basis of project award was always somewhat unknown, continuity of work, personal connections as well as experience being normally the apparent governing factors in the absence of serious fee competition; such competition was, in fact, expressly forbidden by the Association of Consulting Engineers at that time on the grounds that this would lead to a drop in standards. Nevertheless, it was known that clients had not been averse to discussing small abatements of the standard fee scales, where there could be justifiable reasons; the private sector, in particular, being more innovative in this respect.

The whole process had an element of mystery about it; there was a lovely story going around once of one well-known Consulting Engineer, of the newer breed at this time, securing a whole string of small trunk road commissions after taking tea one afternoon in the Ministry of Transport; if true, it was perhaps a lively tea-party!

A further development, not perhaps fully appreciated at the time, was the establishment of Road Construction Units who were charged with implementing the expanding UK roads programme. These were formed from both central government and local authority staff and they grew into large organisations which undertook every aspect of major highway projects; in many instances they effectively prevented Consulting Engineers from participating seriously in the domestic programme, again tending to encourage those firms with these skills to search abroad for their work.

2.1.3. Competition abroad
In the days of the British Empire, much of the work undertaken by Consulting Engineers abroad was conceived in the first instance by government offices in London; this provided the reason for virtually every long-established firm of consultants having their head office, and often their

only office, in Victoria Street, a short step from Whitehall. As the empire changed and newly independent states took control of their own development programmes, so the work had to be obtained abroad in the country concerned. Here, often for the first time, British consultants met international, particularly European, competition. This was very much a new experience and a somewhat chastening one on occasion, when work that was considered the prerogative of a UK firm went elsewhere.

This process really took off with the boom of development in the Middle East during the early 1970s. Here there was, perhaps for the first time, really intense competition that was based not only on the traditional, but somewhat intangible, technical skill values associated with any particular firm of Consulting Engineers, but, most significantly, on fees and on the calibre of staff and presentation skills. In the frenzy of activity throughout the Middle East during this time, traditional fee scales were jettisoned; in any case, such well-defined scales were unknown to much of the international competition and were certainly not recognised by clients in the region!

UK consultancies, however, perceived that they needed this element of overseas work in order to counteract the impact that the cyclical nature of the domestic economy had on their total business, and so many firms, of all disciplines, went for foreign work with a will and with enthusiasm. Before long, virtually all work outside the UK was obtained on a competitive basis; this was the learning place about competition, in all its aspects, for many a budding Consulting Engineer.

Nevertheless, throughout this period, there were not many examples of British firms of consultants suffering financial difficulties. Many of the larger firms still retained their long-standing and inherited markets, whilst the smaller firms remained happy with the relatively stable fee arrangements for the home market. Most firms in the UK, as in many parts of Europe, were partnerships, with a high ratio, perhaps as high as 50:1, of staff to partners. The general rule appeared to be that, as long as a company had sufficient volume going through the firm, then the partners had a reasonable, sometimes a good, living; it didn't appear to be necessary to optimise the organisation and the way it worked to any significant extent.

2.1.4. The UK situation in the early 1980s

That all began to change with the arrival of the new 'Thatcher' administration in the UK, although there had already been signs of what was to come for a number of years.

Work abroad continued at very much its former level, particularly through the UK recession of the early 1980s, but competition between UK and foreign firms and between UK firms themselves intensified as the markets became more developed and as indigenous consultants took an ever larger share, very often at an ever deceasing price. At the same time, the lure of the Middle East faded and it was the Far East that became, for a while, the honeypot for UK consulting engineers; here, however, there was often a well-established local consultancy industry so that UK firms had to adopt a different way of working, which gave them less opportunity to employ resources from within their home offices.

The development emphasis of the international agencies changed from a major project approach to one of sustainable development with its emphasis on learning/teaching and transferring skills, again reducing the ability of UK design offices to contribute significantly to these types of commissions. Increasingly, commissions with these agencies were won on an individual and curriculum vitae(cv) basis, the start of a 'cv sales culture' as one consultant has put it to me; inevitably it meant significantly less work for the large design teams within main offices.

Within the UK, however, changes began to take place which at first sight seemed to favour private firms of Consulting Engineers rather than their public sector counterparts. In 1980/81, the Government offloaded all of the RCUs (Road Construction Units) to existing consultancy firms within the private sector who took over, not only the staff of each unit, but also their commissions. This enabled firms who until now had not really been associated with the roads programme, or indeed with highway work anywhere else, to acquire, at a stroke, a very competent highways engineering capability. This was done on the back of a generally declining workload, thus introducing real competitive pressures for the first time into the heavy civil engineering consultancy market of the UK.

In the private sector, in line with an increasing emphasis on competition generally throughout the economy, competing pressures between firms grew for all forms of private sector development and industrial work. Clients began to take advantage of a serious over-supply situation that was beginning to develop between consultancies, who responded by developing wider portfolios of skills, greater local presence and a more pro-active approach generally in order to win new work. At the same time, the I.T. revolution, the rapid growth of telecommunication networks and the computerisation of all forms of design and management functions, meant that costs fell, work could be undertaken on remote sites, and clients could begin to really get much closer to their consultants and, even, their complete teams of advisers.

This was the period in the UK when all forms of deregulation were first introduced and when central government first began to award consultancy appointments on the basis of competitive fee bids. This was followed, in time, by a general culture of fee competition throughout the whole of the public sector, which even by this time had begun to shrink under the impetus of privatisation policies.

Nevertheless, the old adage still appeared to apply about engineering consultancy, in particular, that if there was sufficient volume of work going through the company, as many of the earlier partnerships had now become, then the partners, or owner Directors, made a reasonable living from the business. Control mechanisms for costs and quality were still fairly rudimentary, and specific 'marketing' or business promotional activities were still somewhat frowned upon in some quarters. It was, however, the start of an even bigger period of transition, and even at this time the signs of what was to come were there, with firms acquiring new disciplines by acquisition, or merging and taking over other firms so as to meet their new strategic objectives.

2.1.5. Summary - engineering consultancies by the mid-1980s

By this time, the wind of change within engineering consultancies had begun to blow, for although there was still generally a good living to be had from consultancy, the competitive pressures were beginning to grow in a serious way, now at home as well as abroad. Changes of many kinds were beginning to be apparent within the stable structure of the industry, not least with the onset of computerisation which now became obtainable at an acceptable price to the bulk of the industry rather than, as hitherto, being the preserve of the larger consultants, or just those who worked on large or prestigious and specialist commissions.

Still, there was a culture of individual rivalry between companies and a secrecy about their own affairs which tended to inhibit most forms of co-operation, other than perhaps on major projects abroad where it was perceived, in many cases, that a multi-national approach would help to secure the project. The instances of "UK Ltd" putting together a single bid for a major overseas project were singularly few and far between, in almost total contrast to our foreign competitors. Clients frequently complained that they had one bid from each of a number of European countries set against perhaps 4 or 5 from the UK; this tended to confuse them and certainly it detracted from the appearance of government support which the UK's rivals were able to capitalise on, to the general disadvantage of the UK and of British consultancy firms individually.

The period of the mid-1970s to early 1980s had provided a 'relatively' stable platform for the engineering consultancy industry; this did not, in retrospect, prepare it well for the 'hurricane' of change during the following 10 years. I believe that, in most instances, many of the older attitudes and characteristics lingered on for far too long, inhibiting change that the market called for and generally not preparing the industry for what simply had to come.

On the performance side, there remained a culture of secrecy about the financial affairs of engineering consultancies, and this still prevails in some quarters even today. Rarely was a firm's financial performance in terms of Profit and Loss, or even rarer, Return on Capital Employed, known in a widespread way amongst the engineering staff; these aspects did not arouse most engineers' individual interests. Such matters were left to be the concerns of the Partners or Directors, who, for their own reasons, tended generally to keep them to themselves.

In the new climate of the mid-late 1980s, consultants now set to, with a will, to compete fiercely and ever more suicidally, on occasion, with one another on the back of a sustained boom in the UK construction industry such as had not been seen for decades. Within many firms, caution was thrown to the winds in the drive to acquire ever more work with limited staff resources; employment costs started to rocket upwards in the face of staff scarcity, standards started to become variable under the increased work-load, and the pattern of gradual and balanced growth, which had been so characteristic of the consulting business in earlier periods, was replaced by a rush to expand dramatically in the domestic market. All this was done in the expectation that the UK's economic performance and growth rate, so long a problem for successive governments, had finally reached a level that would be sustained for years ahead; those who missed out at this stage would surely miss out in perpetuity - this seemed to be the general underlying philosophy at this time.

2.2. MARKET CHANGES IN THE UK
2.2.1. Level of demand
In spite of the apparent upturn in demand in the mid-late 1980s, there was in fact a steady decline in the market for construction as a whole, and thus for the services of Consulting Engineers in particular throughout the 1980s and into the 1990s. *"Overall demand for what we build has declined by 30-40% in recent years and is set to decline a further 20% before bottoming out"* wrote

Sydney Lenssen in 1996.[1] This underlying decline is the background to many of the changes in consultancy over this period; and it is, I believe, set to continue.

2.2.2. Changes in client capability

Inside client bodies, as within engineering consultancies, there was also considerable change and turmoil as a result of the general change in the economic, business and regulatory climate during the 1980s and early 1990s.

The most significant change was the break-up or hiving-off of in-house technical departments, particularly in local government and nationalised industries. There had been a foretaste of this with the disposal of the RCUs in the early 1980s, but now the methods of disposal were much more aggressive. There were 4 main variants:-

- simple reduction of in-house capability by staff redundancy, this capacity being replaced by employing more outside consultants. An example of this was British Coal whose large technical departments were significantly reduced throughout the 1980s.
- sale of in-house technical units, together with their ongoing work-load, to the highest bidder. This approach was used by British Rail as part of its privatisation programme and by the Government's Property Services Agency.
- 'hosting' or short-term offloading of technical departments, with their work-load, to a preferred bidder for a specific period, typically 5 - 7 years, with a possible option of renewal for a further term.
- conversion of existing technical departments into new free-standing consultancies able to compete throughout the market in their own right.

Local government preferred the last two options generally, the latter method introducing new competitors into a market that was already in some turmoil, whilst the former approach raised many problems about the conditions for the transferability of staff and asked the question as to what would be the permanent solution eventually, at the end of the 'hosting' period.

Similar moves within large sections of private industry were made at much the same time; many of these companies had, over the years, developed their own in-house technical departments and they obviously considered that they could streamline their organisations by offloading these support

---

[1] Article in New Civil Engineer, 9th May 1996

departments. They then started to 'buy in' this expertise from outside consultants, from the construction industry as a whole or, indeed, from other professionals, all on an 'as and when required' basis.

With all these changes came an increasing emphasis on reducing the cost of professional services. Client bodies also started to seek to transfer as much project risk from themselves as possible, and there was an expectation that Consulting Engineers would at least pick up some of this. These trends had, for a long time, been present in the private sector, but they now became particularly aggressive in the public sector.

2.2.3. Project requirements

For many years in the UK, there had been a general dissatisfaction by many clients with the way in which the construction industry realised the projects which were asked of it. Often, this was reflected in cost and time over-runs, which the client was somehow simply expected to tolerate, and usually pay for! There also seemed to be a feeling of exasperation at the fragmented state of the industry as a whole and at its apparent adversarial way of behaving, both within project teams and at virtually all stages of the project process.

One way of trying to reduce these problems had been to adopt 'design and build' approaches, a practice already the norm at that time in North America and in many European countries. This tended to offload risk from the client to some extent, exposing, often, Consulting Engineers to a greater degree of risk, but the client was not always well served with this approach and there was no guarantee that the end product would be truly 'fit for purpose'.

Other ways of procuring and implementing projects started to be more common, and these included:-

- one-stop professional involvement where a lead consultant, or project manager, might take on the risk for managing all of the consultancy input to a project. This certainly reduced client exposure to disputes within the professional team.
- development of the 'design and build' concept to make provision for the sharing of project risk across all the parties involved, leading, later, to the development of 'partnering' concepts.
- DBOM (Design, build, operate and maintain) projects, with many other variations on this theme. These were intended to ensure that the whole of the project team took on not only the risk of constructing a new facility, but also the liability for maintaining or running it; in this way, there was an added incentive to ensure that the project was both 'fit for purpose' and had optimum whole-life running costs.

All of these variants bring significant added exposure to Consulting Engineers and a very changed role in project risk and development. Suddenly they, and indeed all of the participants in the project, started to become very aware indeed of the risks, costs and liabilities inherent in projects - not just their own, but those of the other participating firms as well! It has meant a lot of learning of new ideas for staff and businesses.

Within the last few years, these types of concept have been taken even further under the Private Finance Initiative (PFI) of the UK Government. Here complete teams of consultants, contractors, operators and financiers have to be set up to deliver and operate a project over a concession period that is typically 25 years. A large number of projects have now been procured in this way, especially in the health, prison and transport sectors; not only does the consortium for the project now have to carry all construction, operational and maintenance risk, it also has to take financial risk and, often, a degree of revenue or usage risk.

Whilst these new ways of procuring projects have now become commonplace, with the consequent need for the professionals involved to have a much broader range of skills, there has also been a move to offload routine maintenance and facilities management on to companies outside the client's own normal organisation. This has again been particularly prevalent in the highways sector, but it also has been applied across the whole of the government estate, especially defence. This move, also, places completely new demands on the professionals involved, often now Consulting Engineers, in terms of both the level of technology and management and specialist skills.

2.3. OVERSEAS MARKETS

Markets outside the UK were also changing through this period, in both the developed and the developing world.

Consultancy requirements in the developing world changed rapidly to reflect a shift from traditional infrastructure project work, with its high engineering component, to a demand for expertise in project management, project maintenance and human resources management and training. The engineering component of any project could increasingly be undertaken locally. This shift led to the need for staff credentials to be the key element in project award, rather than the reputation of the consultancy as a whole; the demand for appropriately qualified professionals grew, but that then posed the problem of availability and an ageing staff profile of those with the requisite skills. It also meant that the field for commissions became much more open to competition from any international consultants and from other professionals, such as accountants and management consultants, who could simply include one or two key technical and engineering specialists within their teams for perhaps more wide-ranging work.

In the markets of Asia and the Middle East, local expertise became increasingly available and on a par with UK and European consultants; again, then, it was the individual specialist skills that were sought from UK consultants rather than overall project involvement in many cases.

Many of the changes within the UK market that I have described have now also started to take place in Europe, but somewhat behind the UK. This has given some advantage to UK consultants in the short term, since they have already gained appropriate expertise. However, in spite of a number of European Directives following on from the 'Open Market', there has been generally slow progress made in ensuring effective international competition within the EU; partly this can be attributed to language difficulties, where UK professionals are notoriously deficient, but probably it is also a result of natural national defence mechanisms. The European market remains difficult for UK firms to penetrate effectively and in depth, other than within partnering arrangements. The language barrier, of course, does not really exist for European firms wishing to work in the UK, since many of their staff have English as a good second language; however, I feel sure that Dutch and Scandinavian clients, for example, will continue to want all of their work undertaken in their native tongue and this is a big hurdle to overcome for almost anyone outside their particular countries.

When looking further afield, some of these recent changes in the UK and European markets were already accepted practice for certain types of projects in North America and South-east Asia. The impact of change in these places has therefore not been so great, although many similar trends can be identified.

2.4. THE RESPONSE TO MARKET CHANGES

These changes in the various markets forced firms of Consulting Engineers to change the ways in which they had traditionally worked. In general terms, the market has become much less predictable and the changes quite rapid, certainly in the UK since the late 1980s; this has meant that it is those firms that have adapted quickly which have been the more successful. The market world-wide for engineering consultancy is now very much a market that favours the 'fast-footed', so some firms have learnt to become very adaptable to change, with all the implications for staff and resources that such changes have.

Some of the main responses by firms to these market changes are illustrated below:-

- Clients have started to make it clear that they prefer to deal with the 'expert' rather than the manager of any section of the business. As a result, firms have had to employ more experts in a front-line role and increase their range of in-house disciplines.
- To satisfy this demand from clients, firms have also had to improve the local availability of their experts; this is very much one of the reasons that has prompted the spread of offices geographically and the perceived need for firms to be larger so as to support this network effectively across all disciplines.
- Design processes have become increasingly computerised, with more rigid design routines so as to satisfy clients' requirements for quicker delivery of service and a more rapid evaluation of alternative options for a project. Again, this has had staffing implications, with particularly the number of technicians being reduced as engineers undertake more of the work themselves on-line.
- Fees have fallen in response to market and competitive pressures. This has tended to reduce the input that firms have been willing to invest in a project, perhaps also limiting their ability to be creative.
- Companies have become much more professional in their marketing approaches and have had to engage in a much wider dialogue with clients as to their real business needs. Only then can they put the appropriate expert in front of the client who will, hopefully, finalise the brief and, hence, the commission.

All of these aspects have had significant repercussions on the whole industry, both within and outside the UK. Different management styles have started to develop, which separate out the managers and the marketing and business people from the experts and the technical people. Yet all have had to develop a greatly increased commercial awareness and understand that the client's paramount interest is total project out-turn in terms of time, effectiveness and cost. At the same time, cost pressures generally within businesses have led to slimmer organisational structures and reduced waste and overlap of activities.

With the changes that have taken place with in-house technical departments, consultancies have also broadened their disciplines, especially in the maintenance field, as they have acquired former public sector teams. This has secured, in the short term, this particular client base and maintained turnover. In line with a general expansion of the larger consultants, as they have sought to enter new markets, they have also acquired specialist niche consultancies, again to secure a client base and future workload.

Finally - and I would say at long last!- the market pressures have led to an increased willingness by Consulting Engineering companies to collaborate with one another so as to optimise their skills and share costs in the face of a demanding client base. They appear to have recognised that the key need is to put the best team in front of a client, even if each member, by doing this, only secures a share in a project rather than the whole. There seems to be a realisation that effective collaboration can greatly improve the chances of success at the consultant selection stage, and that it may be easier and more beneficial, in the longer term, to secure smaller shares in a number of jobs rather than the whole of a single one. This trend has broken down barriers and secrecy between rival consultant firms, so that they have begun to see that working in combination can secure more work; in addition, a client is often happier and feels less exposed with a joint team of consultants than with a single firm.

This picture is one that is prevalent world-wide, with firms of all nationalities adopting much more a pattern of collaborative working with local or indigenous consultants, in both the developed and developing world, seeking always to form relationships with others that will add value to the project and increase their own credibility.

NOTES ON THE CHAPTER

CHAPTER 3

Engineering consultancy today

In the previous chapter, I have identified and explained some of the tremendous changes that have taken place in the engineering consultancy industry during the last 15-20 years, particularly in the UK; similar changes are taking place, although perhaps somewhat more slowly, in most countries of the EU and in many other parts of the world. It's my belief that even more far reaching changes are going to take place in the next 5-10 years.

First however, it's helpful to look at the general shape and structure of engineering consultancy today using the UK situation as very much a model of how the profession has matured and is poised for further change.

3.1. STRUCTURE OF THE UK INDUSTRY
3.1.1. Information sources

Information sources about the general size and activities of UK Consulting Engineers are fragmented, and there is no comprehensive central register about companies in this sector. In the case of financial performance, the problem is compounded by the ownership pattern that has developed, with many (smaller) practices being partnerships, hence having no published accounts, whilst the larger firms tend to have more complicated ownership structures, with several being subsidiary companies of other UK or international groups; this makes their precise performance quite difficult to discern with any degree of certainty in some cases.

Until a few years ago, there was a reasonably consistent approach to membership of the consultants' trade association, the Association of Consulting Engineers (A.C.E.), but more recently membership of this has become fragmented; the Association, in any case, only kept basic statistics about its members. There are also other trade bodies to whom Consulting

Engineers belong, but membership of these varies according to the emphasis in any single consultant's work and interest. A further complication is that it is the A.C.E. which lays down professional standards for member firms, whilst it is the professional institutions, e.g. the Institution of Civil Engineers, the Chartered Institution of Building Services Engineers, etc. who lay down the professional qualifications for their indivdual members.

For an overview of the engineering consultancy industry, a reasonably appropriate source of data is the Annual Consultants File published by New Civil Engineer.[1] Virtually all civil and structural engineering consultancies tend to contribute to this survey with information about their organisations; most of the larger multi-discipline or specialist practices also participate. The quality of the information is dependent on the firm concerned, but since this particular file is widely regarded as the consultancy handbook, there are strong incentives to get the information provided reasonably correct. This magazine gives a fair snapshot of the affairs of the profession, and its data would appear to correlate quite well with UK statutory data that I have examined. Comparing this source of information with that which is also available from the A.C.E.[2] gives reasonable consistency.

3.1.2. Size of consultancies

The NCE Consultants File for 1999 has entries for 198 firms with staff complements across the range from 20 up to over 9000; of these, almost half have more than 100 staff, whilst the largest 40 companies together employ nearly 70,000 staff, probably around 80% of all practitioners working in the industry once very small or individual consultancies are excluded.

The general spread of company size is shown in Table 3.1 and the corresponding distribution of staff in the industry in Fig. 3.2(opposite).

It's interesting to note that, in 1987, almost the same number of firms, but not the same companies necessarily, had over 100 staff each but the average size of the largest 16 firms then was 1100 whilst today it is some 3400 staff. It is generally the larger firms that have grown very significantly during the last 10 years, with the largest 19 firms, all with over 1000 staff, accounting for almost 2/3rds, some 57,000, of all employment on the NCE list. Even allowing for the inevitable missing entries, mainly amongst the smaller firms, or amongst the handful of larger engineering consultancies which have no activities at all in the civil or structural engineering field, it is

---

[1] New Civil Engineer, Consultants File, April 1999. Published by EMAP
[2] Report of the Forward Study Team. A.C.E. April 1995

clear that the vast majority of employment in the UK engineering consultancy sector lies with the larger firms.

No. of Staff	No. of Companies
More than 2000	10
1000 - 2000	9
500 - 1000	19
300 - 500	15
100 - 300	41
20 - 100	100 - 200
less than 20	several '00's

Table 3.1 *Size of UK Engineering Consultancies - All Staff*
(Source - NCE 1999)

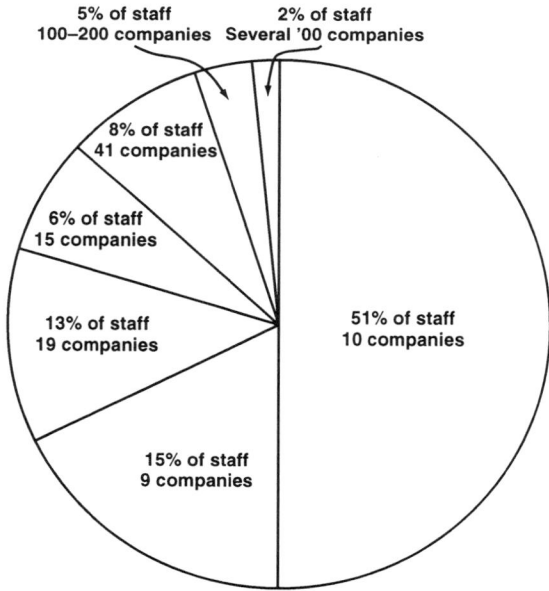

Figure 3.2 *Distribution of staff amongst leading UK Consulting Engineers* *(Source - NCE 1999)*

There are no consistent data available for the precise number of smaller consultancy firms, particularly those with less than 20 staff, but in 1995 the A.C.E. had around 450 member firms in this category and so it is likely that the total number in operation in the UK runs into several hundreds and probably into thousands; many of these will be sole practitioners. Thus, the picture of the industry is one that is generally fragmented.

With such disparity across the industry, there will be few common concerns between the very large and the very small consulting engineering firms, certainly as regards client base, work profile and, maybe, personal life styles. However, all are likely to share the same business concerns of:-

• finance
• marketing
• resource availablity and planning
• quality issues
• technical and production problems
• variability in the construction market.

3.1.3. Ownership

Of the largest 50 companies on the Consultants File, some 30% are subsidiary companies of parent groups of one form or another, whilst 20% are owned by foreign consultants, many of whom originate in the U.S.A., alongside a smattering of European groups. Most of the remainder are independently owned, either by the partners (if still partnerships) or by the Directors and senior staff or, in a few cases, within some form of Trust/Foundation arrangement.

This ownership pattern is very much in contrast to the situation only 10 years ago when almost all the firms in the sector were in their own ownership.

Again, as the size of the firm gets smaller, the percentage of firms being owned elsewhere falls and the traditional partnership structure is predominant.

Eight companies are public companies, most of whom are quoted on the UK stock market, but often with a significant staff shareholding, and more are probably considering this route as a way of gaining improved financial flexibility; on the whole, there is otherwise little evidence of commercial companies such as banks, conglomerates, etc. wishing to take stakes in the sector, due, presumably, to its generally poor and certainly variable financial performance. Most outside ownership arrangements have a logical or sectoral rationale behind the investment.

3.1.4. Spread of offices

During the last 15 years, consulting engineering companies have increased considerably the number of offices from which they operate, although many have started to reduce these in more recent times; this applies both to their UK operations and to offices abroad. It is a reflection of their wish to have a more local presence in the market and of the need to be able to demonstrate to local clients that they have a man, or woman, 'around the corner'.

Even in the case of UK national clients, e.g. the Ministry of Defence and the Department of the Environment and Regions, whereas 10 years ago they would select for a project off a national list, now, following the decentralisation of many of their activities, they behave much more as a genuine local client and expect to have ready access to an office that is reasonably convenient.

Companies which employ more than 1000 staff will typically have around 15 UK offices, the range being from 10 to 25, in the case of one of the largest firms; overseas, UK consultants of this size have typically 20 or more offices in all, although in some firms the number of declared overseas offices includes offices staffed by other parts of their parent group. This number probably also includes some offices run in practice by local indigenous firms with whom the UK consultancy has an association arrangement for the occasional project.

The number of offices reduces as the total size of the firm is smaller; firms with a payroll size of 500 staff might have 10 UK offices with some offices abroad whilst firms with a size of 100 staff might typically operate from 3 or 4 offices. A.C.E. data[3] indicate that, of their members, over 50% operate from a single office; these will be the small firms or sole practitioners.

3.1.5. Management structures and systems

The large number of active offices which most of the larger firms operate gives an idea of the complexity of their operation in terms of management, cohesion and control. An additional factor, and one of possibly even greater significance, is the fact that most of the larger firms trade on a multi-discipline basis, and so each office will have a range, if not the total spread, of engineering and other disciplines; this reflects the perceived need to have

---

[3] A Future for A.C.E. A Report. A.C.E. October 1995

reasonably recognisable expertise available locally, certainly for the key areas of work. Few of the larger consultants appear to run networks of single discipline offices, at home certainly, with the possible exception of the large mechanical and electrical engineering practices, and even abroad many of the offices have a significant range of disciplines.

For management, this dispersion of multi-discipline offices poses significant challenges in organising the activities of the company in terms of:-

- project performance, including scheduling, quality, costs, resources
- financial performance and cash management
- marketing, business promotion and development
- future planning strategies.

All of these need to reflect both the locational aspects of the different offices as well as their particular skills and resource availability.

There appears to be no real consistency of approach between different firms as to how these issues are managed; a variety of management structures are used, ranging from full matrix systems which tend to be very complex, to total delegation of virtually every aspect of the business to individual operating units. This latter approach can lead, ultimately, to a situation where separate units become almost totally autonomous operations which can begin to question whether they have, or indeed wish to have, a meaningful role within the total company or not.

So far as management systems are concerned, these are now always computer-based with, in some instances, sophisticated real-time and paperless reporting. Offices are networked so that detailed reports at both project and operating unit level can often be obtained remotely. The level of sophistication of some of the systems in use has surprised me; one wonders whether such systems are really necessary for this industry and whether the costs of implementing and running them represent good value for money.

The management of all these types of issues is certainly going to be a very real challenge in the years ahead, if not already a thorny problem; the way in which they are resolved is probably going to impact very significantly on a company's financial performance - a subject to which I now turn.

3.2. PERFORMANCE OF THE UK INDUSTRY

For the reasons which I have already set out, it is quite difficult to obtain reliable data on the performance of the consulting engineering sector as a whole, although one can always obtain reported figures for individual companies, but obviously not for partnerships. However, it is possible to make a general and overall assessment on the performance of the sector using the information that is published about individual companies.

3.2.1. Turnover

This is the most robust measure of the level of activity for any consultancy firm; it can be examined in accounts and is generally reported. Table 3.2 summarises the turnover level achieved per member of staff and typical turnovers for companies of diffferent sizes using average figures for turnover/employee.

Turnover / Employee - £'000 p.a.	
Range	42 to 65
Average	52

Average Turnover of Consultancies - £'M p.a.	
Staff Nos.	**Turnover**
500	26
1000	52
1500	78
2000	104
3000	156

Table 3.3 *Turnover vs. staff numbers*

These figures indicate the variability that occurs across the industry, and make interesting comparison with those from earlier years because in the last 4 years there appears to have been a very slight rise generally; for the 5 years before that, the figures were stable. Thus, although salary costs per staff member have risen with inflation, fee generation has not to the same extent, suggesting, at first sight, that:-

• profitability has fallen, or
• companies have got greater control of their overheads and ancillary costs, or
• clients are benefitting from lower unit fees.

Yet, as shown in Chapter 2, clients have generally been demanding more input for their fees, so that there must also have been an element of increased computerisation and/or efficiency combined with, possibly, a lower level skill input for the turnover generated.

3.2.2. Profitability

It's in this area of company performance that the traditional secrecy of UK Consulting Engineers sometimes prevents full and open disclosure. This is, I suspect, a legacy from the old partnership days, for in many companies the former Partners are now the owner Directors. Many of the company structures are complex, and true profitability figures for trading groups as a whole are difficult to derive, let alone for separate discipline or geographical subsidiaries.

Appendix 1 has a detailed analysis of a sample of consultancies; I have selected these to try to represent a good spread of companies across the sector, for size, activities and interests. Table 3.3 summarises some of the performance figures for those companies that are not subsidiary companies of a parent group.

PROFITABILITY %		
Non Plc's	Range	0 to 5
	Average	2.5
Plc's	Range	4.5 to 18
	Average	8
RETURN ON CAPITAL EMPLOYED % (ROCE)		
Non Plc's	Range	0 to 26
	Average	13.3
Plc's	Range	16 to 62
	Average	37

Table 3.4 *Summary of Financial Performance*

The Table shows that there is, indeed, a lot of variability, even with the loss-making companies taken out of the analysis, but, generally, a fairly dismal performance, certainly from the point of view of any investor. The amount of work overseas *might* be a contributory factor, as will the range of disciplines and type of client base for the company concerned. Current profit levels for the sector can be compared to those for 1992/3, where the average profit level for larger UK consultancies was 2.5%.[4]

It's noticeable that those consultancies which are plc's have very significantly higher profitability and ROCE figures. This must reflect their enhanced financial control and awareness together with a need to demonstrate shareholder value so far as possible; nevertheless, the returns to shareholders can hardly be exciting. There is little correlation between these higher overall performance figures and the corresponding turnover/employee figures, which are little different from the run of other, non plc, companies.

It's also significant that the performance of those consultancies which are subsidiary companies appears to be generally poorer than those which are free-standing consultancies in their own right; indeed several are recently reported as being loss-making. This *may* be due to variations in intra-company trading and cross-charging, etc. and this makes it difficult for outsiders to assess the precise performance of this component of the group on a free-standing basis; there may also be transfer pricing and overhead recovery issues which may be affecting the profitability of the subsidiary. Tolerance of this state of affairs may also reflect the fact that the parent company might perceive the value of the subsidiary more in terms of its market development potential for the remainder of the group rather than simply in terms of its short-term profitability.

The industry itself appears to share this general concern about overall levels of financial performance, with the recently appointed Chairman of Halcrow Group, 65% of whose work in 1998 came from overseas projects, saying *"It is a fact that we are working in an industry that does not deliver high enough margins."*[5]

Industry comparisons can always be dangerous, especially when the sector that engineering consultancy belongs to - *'Architectural and engineering activities and related technical consultancy'* - is a very large one comprising a wide range of very different technically-based companies. However, median and quartile figures for the sector as a whole are indicated in Table 3.5; even the best-performing engineering consultancies score poorly by these comparisons in investment terms, although the high level of turnover per employee must indicate that many of the other companies in the sector

---

[4] Internal company research by DHV
[5] Derek Pollock, as reported in N.C.E., 29th April 1999

are either operating on a much lower level of technology or are achieving their results with much lower overheads, perhaps on a secondment or outsourced basis primarily. Clearly the cost and fee structures associated with engineering consultancy in the UK are such that overall financial performance is, at best, average.

	Lower Quartile	Median	Upper Quartile
Return on Capital Employed(%)	32	120	500
Profit/Turnover(%)	5	30	57
Turnover/Employee - £'000	19	34	66

Table 3.5 *Industrial sector comparisons*[6]

3.2.3. Cash management

It is difficult to compare the efficiency of the cash management process between different companies in the industry due to the large differences in reporting and financial procedures, especially when looking at the smaller free-standing companies, partnerships or subsidiary companies. A brief overview of the financial returns does, however, suggest that cash management for the industry is a perennial problem, with typical work-in-progress indicators being around 10-15% of turnover, and debtors being collectable at 2-4 months, a very high figure for any service-based industry.

3.3. DISCIPLINES OF UK CONSULTING ENGINEERS

The NCE Consultants File also identifies the split between 'all staff' and 'civil and structural' engineering staff from those firms who are entered on the file. This indicates that amongst the larger companies, say down to 500 employees, up to 50-60% of technical staff were primarily civil and/or structural engineers; this proportion then rises progressively as the size of the

---

[6] Juniper Overview Reports. ICC Information Ltd, Aug. 1999

firm decreases, so that those firms with a total staff complement of 100 might have 75% of these particular disciplines. Generally this proportion rises even further as the size of firm gets smaller still.

It's important to note, at this point, that several large and well-known firms of mechanical and electrical engineering consultancies, in particular, do not feature in these lists. The reason for this is that several of these firms confine their activities to their specialist disciplines and have not widened out their range of skills to anything like the same extent as their civil and structural counterparts have done; they therefore do not feature in the lists produced by New Civil Engineer and there is no comparable source of information for them.

3.4. RECENT TRENDS IN THE UK INDUSTRY
3.4.1. Performance

Comparative figures suggest that the UK has a high proportion of engineers described as engineering consultants within its general population in comparison with anywhere in Europe. Given that the UK construction and project market is certainly no greater, proportionately, than that of other countries, and is probably a good deal smaller than many, the need for UK engineering consultants to continue to work extensively outside their home territory is obvious.

Competition throughout the world market is ever increasing and new competitors are being created all the time; thus there is a perpetual underlying pressure for UK consultants to increase their share of the home market, wherever possible, at the expense of their direct competition. Consultants have done this by ever widening their disciplines and by undertaking new forms of work, e.g. maintenance, which might previously have been the preserve of in-house or public sector personnel. This is perceived as enabling them to retain client loyalty and to continue to cover fixed costs, but it is probably often at the expense of margins.

The fact that turnover/head has not really increased over the last few years is very significant; my own experience in this time was that people had to work very much harder and have a higher output. This must confirm that fee levels have indeed continued to fall for each commission. Yet, the market is tending to demand that consultants take larger risks on projects, both prior to award and during execution. For many companies, this inhibits not so much their ability to bid for work, although there are signs of greater selectivity here, but it certainly inhibits their ability to have any significant stake in the project. The financial soundness of most companies is simply not there.

The continued widening of the geographical and discipline base is unlikely to be the way of resolving these difficulties within the UK market. Added to this, there is the problem of fragmentation in some of the larger organisations, where competing networks can develop internally, especially when separate operating units are self-accounting. This can have a devastating impact on financial performance. All of these factors combine to paint a picture that could be fairly bleak for UK Consulting Engineers, given a continuation of current trends.

Clearly, there is much to address within this industry, and fundamental change is probably required for many companies, covering organisational, cultural and financial fields. In many ways, a lot of these changes are required and are being implemented in other professional service industries which are in equally competitive markets. I look at some of these aspects in Chapter 4.

3.4.2. Client and employee satisfaction
In recent years, the UK construction industry as a whole has come under the spotlight of the UK Government for its seemingly woeful performance. It is perceived as an industry riddled with conflicts of interest and adversarial practices, and this has led to a number of reports and initiatives being commissioned which have looked at new ways of undertaking projects and thus achieving higher client satisfaction and cost reductions. These have included:-

- The Latham Report - 'Constructing the Team'
- The Egan Report - 'Rethinking Construction'
- The Construction Round Table
- The Movement for Innovation (M^{4}I)

So far as Consulting Engineers are concerned, they are very much part of the overall process and so they are included within these overviews. However, they are themselves looking more closely at their own activities and have commissioned reports through the ACE on the future of their profession. Perhaps one of the more telling condemnations to come from these activities has been the perceived need to publish a 'Client Guide' at the end of 1998; this guide was produced to *'help Consulting Engineers to understand better their smaller and one-off clients'*. It seems to me that this is a damning comment for any service profession; surely members of the profession should know this well enough?

On a more personal level, it would appear that civil engineers, across the whole of construction, are far from happy with their own jobs and prospects. In a recent survey done by Bath University of 138 vocations and professions, civil engineering came near to the bottom of the table in *'happiness'* stakes. Factors in this would seem to be:-

- Inability to work flexibly and on their own initiative
- Lack of understanding by the general public of the civil engineer's role in society
- Conservative nature of the profession, and its slow prospects
- Lack of training
- The small number of large projects in the UK and the frequent political frustrations associated with these
- Concentration on reducing cost throughout the construction process, rather than achieving optimum projects.[7]

All these factors suggest an industry and profession that, in the UK at least, is not at ease with itself and is very much in need of new initiatives.

3.5. ENGINEERING CONSULTANCY ELSEWHERE

Many of the trends which I have observed in the UK industry are repeated in the engineering consultancy industry in most developed countries. The extent to which these trends force the larger consultants to seek work outside their home territory is, generally, much reduced in comparison with the UK situation, with perhaps the industries in Holland and in Scandinavia being notable exceptions. Often the sheer size of the domestic market, e.g. in the USA or in Germany, is sufficient to provide an adequate home-base for most firms, including the larger ones, and this will account for many of these companies appearing to be less enthusiastic, generally, about working overseas.

3.5.1. Information sources

Information sources about engineering consultancy outside the UK are also more difficult to access and do not appear to be as comprehensive as in the UK. I have not found much national data which is comparable and as comprehensive.

---

[7]Report on the survey by Bath University, NCE, 9th September 1999

FIDIC, the International Federation of Consulting Engineers, is one source of data internationally, but this is limited to those companies that elect to belong to it because they have an interest in working abroad. On each national market there is also the equivalent of the ACE in the UK, which can provide some limited data about its particular members. Otherwise, published literature, that is readily available, tends to be confined to descriptions of project services offered.

The best information on company performance, rather than just on range of services, etc., is to be had by accessing company records directly. I therefore approached a sample group of larger Consulting Engineers within some of the more developed markets so as to get a brief overview. The response proved interesting, to say the least! I received a high response rate, often providing helpful additional information, from the European consultants whom I had approached, but from North America and elsewhere the response was poor. Appendix 2 lists some of the specific performance details of individual companies obtained in this way.

Using these various sources, supplemented by other interviews and enquiries, it is possible to build up a general picture of engineering consultancy elsewhere. It would appear that the overall performance of engineering consultancies outside the UK is not significantly different from their UK counterparts, suggesting that these companies are operating under many of the same pressures.

3.5.2. Europe

The development of firms of Consulting Engineers in Europe has been fairly similar to the way in which firms have developed in the UK, but with specific national differences in a few countries, e.g. France, where the construction industry works in quite a different way to most other parts of Europe. Thus, there are, generally, many large companies of Consulting Engineers, by international standards, as well as a multiplicity of smaller/medium-sized firms operating primarily in their domestic and/or local markets. This fragmented pattern of the industry is very similar to that found in the UK.

I found that the majority of firms whom I approached, responded readily to enquiries, indeed most were positively keen to provide information on their people, their services and their performance. In addition to providing information on their activities and the services which they offer, most publish an Annual Report to supplement their Statutory Accounts. These reports contained the usual review of the year, prospects for the future and descriptions of projects; however, many also made some positive recognition of the intellectual asset base of their staff and commented on likely changes in the industry in the longer term, and how they might be addressing these. This was all quite in contrast to the UK/North American situation where the equivalent sources for many companies tended to be either straight-forward brochure 'selling' material or Annual Reports in the format of large UK plc's, with a concentration on 'numbers' and financial ratios, etc. rather than on underlying philosophies.

Using the figures in Appendix 2, some of the noticeable aspects of the performance of these consultants, compared to their UK counterparts, are that:-

- turnover per head is much more consistent from this sample, although similar to that for the UK at present exchange rates
- profitability is variable and generally low - similar to the UK
- ownership structure for the larger firms is more often of the 'Foundation' nature, or else with significant employee ownership amongst their own national staff
- most have significant proportions of their turnover outside their home territory.

Many of the engineering consultancies in Europe have developed in a similar way to their UK counterparts, but often within a more stable, cohesive and less competitive domestic market, where long-term relationships have been nurtured and have been important factors in continuity of work. Some of these relationships have been very long-term indeed but this is now beginning to change as markets are becoming more open, thus exposing companies to greater competition, both from domestic and foreign competitors. The result is often a search to broaden activities and geographical spread - just the effects observed over many years in the UK situation.

My own experience of working with European consultancies is that there is a greater open-ness and willingness to share information, problems and aspirations that is not often found elsewhere, certainly not in the UK industry. There is also, certainly within some countries like the Netherlands, almost an expectation of jointly sharing large projects with competitors in joint-venture or within some national umbrella organisation like, for example, NEDECO, which is formed of a group of leading Dutch consultancies for the sole purpose of working elsewhere in the world on a specific project by project basis. Various Dutch Government organisations also participate in NEDECO on projects for which their in-house specialist expertise is required; this helps to provide an additional level of credibility to the project team and is a pattern that is repeated in a number of other European countries for work outside Europe. National academic institutions are also commonly involved in project participation for many of the same reasons.

Many of the larger European firms are of a similar size to their UK counterparts, with Dutch and Danish consultancies, in particular, being large and numerous for the relative size of their own domestic market. Firms of Consulting Engineers from these, and indeed from many European countries, operate world-wide in a significant way; some 28 of the world's largest 50 international engineering consultancies are based in Europe (including the UK), with their main export turnover coming from projects in other European countries, Asia, the USA and the Middle East.[8]

Many of the larger European consultants, as already mentioned, have an ownership structure that is of a 'Foundation' nature, often with extensive employee participation. This is very much in line with a culture of openness and involvement about the future of the consultancy and with a wish to ensure that staff are committed to the company's aspirations. All of these aspects help to ensure that the company can remain independent, free from outside interests, and, to a large extent, protected from acquisition or take-over. This is all quite a contrast to the situation in the UK where there has been much turmoil and changes in ownership amongst leading firms of Consulting Engineers over the last 10-15 years.

All these aspects suggest to me that it is some of the larger European companies that are beginning to grapple in a meaningful way with some of the issues that I raise later in this book. UK Consulting Engineers should perhaps 'beware' of this, for these firms are formidable competitors in the world market.

3.5.3. North America

My research into the consulting engineering industry in North America suggests that many of the trends found in the UK industry are also being experienced there. Thus, although there is a much larger number of firms altogether, and some of these are very large indeed by international standards, there is still an enormous number of small consultancies, mainly in private ownership, operating in very local markets; as these smaller firms have grown, either organically or by merger, so they have encountered just the same problems that I have explained in relation to UK consultants. The

---

[8] The Top 200 International Design Firms, Engineering News Record, July 1999. Published by McGraw Hill

large consulting engineering companies also have similar ranges of activities and interests that make effective management quite a complex issue.

The whole nature of the industry in North America, as in Europe, is highly fragmented. A lot of consolidation of the industry is currently prevalent, much as in the UK, with several large companies being particularly active in taking over other competitors in order to try and broaden their geographical and client base and their skill mix, and to enable them to invest in new technologies on a larger scale. Thus, there is probably very much a feeling of insecurity amongst medium- sized firms of Consulting Engineers in North America who are well established in their local markets and who are perhaps on the 'target list' of potential larger rivals seeking to improve their market penetration. As in the UK, there is a widening gap within the whole profession between the large players and the small, often highly specialist, engineering consultancies.

Ownership structures, certainly in the USA, are very varied with larger companies in the sector being often 'Common Stock' companies, but with significant staff participation, sometimes world-wide if the company is structured that way. It also appears that in a number of companies, it is shareholder value that is a key issue in performance rather than dividend policy, such that profits are often wholly retained in the business for its development. For smaller companies, private ownership seems to be the predominant pattern, as in the UK and Europe.

Many of the very large 'engineers' or 'design firms' in the USA are in fact 'engineer-contracting' organisations which take on both the design and engineering element of a project as well as some of the contracting work, normally the project management or lead-contractor role. Thus in the annual survey published by Engineering News Record (ENR)[9], of the top 20 companies globally by turnover:-

- more than half are engineer-contractors,
- almost all of these are based in the USA, and
- most of these have very significant or predominant work-loads in the industrial and petro-chemical sectors.

In fact, when it comes to work outside North America, it is almost solely in this latter sector in which the activities of these large Engineer-

---

[9] The Top 150 Global Design Firms, Engineering News Record, July 1999. Published by McGraw Hill

38

Contractor companies is found. The size of projects that these companies are involved in is very large, even by global standards. Thus, the issues that confront these very large companies, with their wide range of disciplines and activities, although similar in some ways to those that equally confront the pure professional consultancies, are different, since the contracting element of their work does make for quite a different mix of people and, probably, for a different approach to management and organisational structure.

Looking at the companies in this list that are purely Consulting Engineers, some 10 out of the top 20, or 26 out of the top 50, are North American-based with annual turnovers in the range US$175m.-1016m., thus whilst some are indeed larger than their European or Japanese competitors, there is not a vast disparity in scale amongst the very large engineering consultants worldwide.

Although it is the North American firms that very much dominate this list in the sheer numbers of companies and in their large overall size, when it comes to turnover abroad, USA firms only generate in total a little more outside their domestic market than the European firms, and, as I have said before, much of this work comes from a limited number of very large projects. In contrast, European firms obtain their turnover by being involved in a very much larger number of projects altogether. By making allowance for the significant volume of work that comes under the category of 'engineer-contracting', it seems that European consultants, generally, are much more active outside their home markets than their North American competitors.

There are also examples now of UK-based Consulting Engineers starting to penetrate the American market, so clearly British firms see a potential for work across the Atlantic.

Many North American Consulting Engineering companies appear to have the same broad range of disciplines as their UK counterparts, although the proportion of their work that is abroad is generally smaller. However, there is probably a higher proportion of their work in the environmental field than might be found in consultants from other countries; this reflects the recognition amongst North American client bodies that the environmental effects of their developments and operations can have significant impacts on their commercial and business performance. Hence, this aspect of consultancy has, for many years, been taken very seriously.

Company performance details appear to be scant, perhaps reflecting a culture of secrecy, but one cannot be sure on this. The ENR list suggests that overall profitability levels for all top international firms are just below 6% of turnover - a figure similar to that found in the UK and in Europe, thus suggesting that overall performance in North America by itself is probably not very different.

What is noticeable is that several large American firms, in particular, have been showing an increasing interest in Europe over the last 10 years. Several of them have made moves to acquire UK consultancies or at least to set up potential joint ventures or partnering arrangements. This appears to have been in order to gain access not only to the UK market, but probably more to the European market in the longer term. Notable examples are Black & Veatch, LAW, Parsons Brinkerhoff and Montgomery. Where such acquisitions have been made, it seems to be a general practice that it is the UK component of the enlarged group which often takes the lead in most of the international operations outside the parent's domestic market.

It seems that the industry, as a whole, like its UK counterpart, suffers from a lack of public perception, exposure and confidence. In a recent comment, Valentine Lehr, President of the New York Association of Consulting Engineers, highlighted that:-

- the public is increasingly distrustful of all professions in general
- the reluctance of Consulting Engineers to take on risk, for them to appear to always be avoiding liability and to avoid taking on controversial tasks, is opening the way for others (*"Engineers make projects run, but do not run them"*, he said)
- civil engineers tend to be analysts rather than synthesisers, i.e. they are the 'worker bees', rather than the initiators and controllers, and,
- poor communication is a recurring theme for the profession.

These comments certainly suggest that much American experience is similar to what is being experienced in the UK. Valentine Lehr concluded that *"there are plenty of clients out there, but you need to talk to them."*

So far as management style is concerned, engineering companies in the USA appear to operate in a more relaxed way than their UK counterparts, although whether this is simply a superficial impression is difficult to say as there are clear variations from company to company. As part of this culture, there is heavy emphasis on the 'empowerment' of each indivdual and on a team approach to work within the company. Recruitment and retention of staff of the highest calibre is difficult across the continent, much as in the UK and some parts of Europe, so this also exercises the minds of management to a significant extent and is high on their agendas.

3.5.4. Other countries

The other parts of the world where engineering consultancy is well-developed are the Far East and Asia, Australasia, South Africa and South America. The patterns of development for the industry in these regions have been quite similar to European experience. The majority of these engineering consultancies tend to be fairly domestic, although many are large firms in their own right; a possible exception is the case of the larger Japanese consultancies who do export, primarily to elsewhere in Asia, but also, to a limited extent, to other parts of the world particularly to the Middle East. Some of these companies are also described as Engineering Contractors in the ENR tables.

It would seem that the domestic markets for most of these companies are fairly well-protected, thus providing little incentive for them to search outside them, but firms in these areas appear, in the longer term, to face similar threats and difficulties to those facing UK and European consultants.

3.6. OTHER COMMENTS

Overall, it would seem that engineering consultancies in many developed countries are facing many of the same kinds of difficulties as the industry in the UK. The common themes coming out of all these sources are:-

- the importance of communicating with, and relating to, clients and the general public
- the need to value staff and their skills and thus achieve effective staff participation in the business
- new ways of assessing company performance overall and ensuring long-term viability
- building new relationships and alliances everywhere
- working in new ways to suit the IT age
- embarking on new ways of being involved in projects
- ongoing training for all
- broadening the skill and client bases through the acquisition of, or merger with, competitors.
- partnering and alliances are seen very often as the way forward for many companies trying to be involved in large projects
- there is an inceasing tendency for companies to have to take a financial stake in large projects
- profitability levels, almost everywhere, and certainly amongst international competitor consultants, are generally modest.

Some typical comments from published reports are given in Table 3.6. These illustrate some common concerns and approaches from a variety of consultancies of different nationalities.

Table 3.6 *Some significant comments selected from Annual Reports provided by Companies*

Atkins (UK)	An important part of our planning continues to be the expansion of our outsourcing business(54% of current turnover).
Arcadis (NL)	Need to acquire companies to achieve growth and to improve our strength in certain sectors.
Carl Bro (DK)	An 'Intellectual Capital Account' to assess whether we have managed to provide an appropriate platform for future earnings. 43% of staff have the option of teleworking. A policy to ensure the right balance of age, creativity and experience across the group.
COWI (DK)	A 'Knowledge Account' to reflect skill levels and customer satisfaction.
DHV (NL)	Alliances are needed to spread risks and to provide all the required project expertise.
Fichtner (G)	Need to partner to access some markets. Taking project risk can improve returns.
Haskoning (NL)	High staff turnover reflects labour shortages. Need to instil more of a spirit of 'belonging' and to instigate management training.
High-Point Rendel (UK)	The focus of our business has continued to move towards the provision of business and management consultancy services. These now account for 64% of the profit on 41% of turnover.

Hyder (UK)	Our engineering and environmental consultancy business performs in often difficult and highly competitive markets. Very significant involvement in PFI work.
Holland Railconsult (NL)	In the future, the pattern will be of small organisations working together for a project. As 'virtual environments' become more usual, no need to be a specialist to enter and access a market.
LawGibb (US)	...greater concentration on growing the....revenues derived from existing clients will enable us to capitalise on these relationships profitably.
Norconsult (NOR)	Around 3% of our turnover was allocated to training initiatives and to R. & D.
Oscar Faber (UK)	Concept of a 'business scorecard' to reflect business relationship, client, staff, supplier and shareholder interests.
Ramboll (DK)	Our Holistic Operations lead to the concept of Holistic Accounting, reflecting both business and staff issues.
RPS (UK)	The late 1990s have seen corporate recognition that environmental issues are capable of affecting revenues, costs and asset values.
Stantec (CAN)	We continued to pursue consolidation in our industry by successfully completing 9 acquisitions. The company will continue to grow by acquisition in our highly fragmented industry.
Waterman (UK)	The acquisition of CPM will enable us to raise Waterman's profile... and will increase the opportunities for cross-selling the Group's services.

White Young
Green (UK)   The quality of earnings has been enhanced through
                         the changing nature of our relationships with
                         clients and the length and type of contracts
                         secured.

Witteveen+Bos (NL) Average age of staff 35 - emphasis on youth and
                         creativity. All profit distributed to employee
                         shareholders.

WSP (UK)   Partnering and term appointments now represent
                         more than 50% of our existing business. Closer
                         working between consultant and contractor will
                         lead to more single-source procurement.

NOTES ON THE CHAPTER

CHAPTER 4

Changes in working cultures - Work and organisations in the future

4.1. INTRODUCTION

Much has been written about the dramatic changes in working patterns that have taken place over the last 15 years throughout the (former) western world; it is useful to pick out some of the thrusts of these changes and to see how they might continue in the future and, in turn, become relevant to the businesses of Consulting Engineers.

In the world of work in its widest sense, perhaps three examples are a good starting point:-

"Already an estimated two-thirds of U.S. employees work in the services sector, and 'knowledge' is becoming our most important product"[1]

" Less than half the workforce in the industrial world will be holding conventional full-time jobs in organisations by the beginning of the 21st century. Those full-timers or insiders will be the new minority"[2] Already, in the UK, in 1993, only 55% of the adult work-force which wanted to work was employed full-time, and the figure is falling rapidly, one suspects.

" Constant training, retraining, job-hopping, and even career-hopping, will become the norm".[3]

---

[1] Post-Capitalist Society. Peter F. Drucker. Butterworth-Heinemann

[2] The Age of Unreason. Charles Handy. Arrow, 1989

[3] Global Work: Bridging Distance, Culture and Time. Mary O'Hara-Devereux and Robert Johansen. See Notes to References

And then there is the classic prophecy of the future factory:-

" The factory of the future will have only two employees, a man and a dog. The man will be there to feed the dog. The dog will be there to keep the man from touching the equipment."[4] In this example, whilst a sombre outlook for manufacturing employment, there is at least some good news for Consulting Engineers. The factory and its process equipment has to be planned, designed, managed and constructed in the first place, and there will always be an ongoing facilities management and maintenance requirement which consultants might take charge of.

It would be easy for us to say that these examples and prophecies really affect only the industrial or manufacturing part of the economy, rather than the purer service sector; however these trends are already discernible in many service industries, and so it must only be a matter of time until they become equally relevant to engineering consultancy.

This chapter looks at future trends and scenarios under 3 headings:-

- Organisational and Management Structures
- Working Methods
- Employment Arrangements

4.2. ORGANISATIONAL AND MANAGEMENT STRUCTURES

4.2.1. Downsizing, the core worker concept

During the last 15 years almost all organisations have reduced the staff complements of their businesses either by physical reduction of staff numbers or by the creation of separate operating units. More often it has been by a combination of both of these methods with the business in question being more strongly focused on its main activity and peripheral activities being closed or hived off elsewhere. In particular, the corporate centres of major international companies have been dramatically reduced in this way as the whole organisation has been broken into its component parts, leaving only relatively few key functions to be carried out in the centre.

This rationalising of activity has led, inevitably, to a reduced requirement for staff to be employed full-time by the business so that those who are left can be referred to as 'core workers' or those employees who are indispensable and essential for the carrying out of the core activities of the business.

---

[4] Warren Bennis, Professor of Business Administration, University of Southern California. See Notes to References

Another component of this change has been the concept of further reducing permanent staff by outsourcing a whole range of non-core activities; facilities management, security, even word-processing, spring to mind as typical examples of the kind of support services that can be put out. These services may not be required at a constant level all the time and can, therefore, be more economically hired in as and when needed; alternatively they might be more effectively provided and managed by experts in their own field rather than by mainstream (i.e. core) staff who should not have their attention distracted from the main purpose of the business. Thus the plant manager can concentrate on his main function, i.e. factory production, and the project engineer on managing his/her projects without the distraction of peripheral administrative activities for which neither has, normally, been trained at all.

A recent example of this type of pattern was the news that Scottish Widows, a major life assurance company, was contracting out all the administration of its £22bn. portfolio to WM Company, the Banker's Trust subsidiary. This administration is reckoned to cost £60,000-£120,000 per head of staff, so the company obviously intended to make significant savings in this way.[5]

This focusing of management attention on the main purpose and objectives of the business is the driving force behind these changes. Key managers are primarily concerned with strategy, watching every move that their competitors make in our competitive world today, and constantly assessing their company's position in the market; this attention cannot afford to be spread too widely or be distracted by 'administration' issues, as then its focused effect is lost.

One of the consequences of these very much leaner organisations is that people are stretched more - *"stretch them thin"* as Rosabeth Moss Kanter has called it.[6] It has the effect of making the people who remain in these parts of the organisation devote more and more of their time and personal life to their work. Charles Handy in 'The Empty Raincoat'[7] describes it in another way as *"half as many staff working twice as hard and producing three times as much!"*. It is very symptomatic of our times, and, where it has been applied, it *should* have enabled companies to both gain competitiveness and increase volume. Often this has not appeared to be the case and the opportunity to increase

---

[5] Financial Times, 9th February 1996
[6] When Giants Learn to Dance. R.Moss Kanter. Routledge, 1989
[7] The Empty Raincoat. Charles Handy. Arrow, 1995

volume without additional cost has not been grasped; as time goes by, the company concerned has then found itself unable to increase its levels of activity due to shortage of in-house personnel!

As this process continues, more people find themselves displaced from the old 'central' team into careers in 'producer services' such as advertising, data processing, personnel supply, management and business consulting, security, maintenance, legal, accounting, etc.,etc. The list is almost endless, with many sectors also putting out their professional engineers and architects, sometimes by way of an arrangement with 'host' companies of consultants who thus take over the workload risk of these staff.

As shown above, one of the other by-products of this process is that the remaining 'management' or 'central' team for the organisation has little capacity to take on any new initiatives or deal with unexpected, but perhaps urgent, tasks. This leads to the need for outsiders to be brought in, perhaps for a specific task - a field of opportunity for management consultants and interim managers; this has the effect of opening up the organisation to new ways of thinking and of working with people from outside, on a regular basis.

4.2.2. Flat management structures and restructuring

We live in an age where 'quality' is very much a buzz word, but it covers the general view that clients are becoming more demanding in their search for a consistent approach to the quality of work and products; for service industries, the word does not just refer to technical content but to the whole way that a service is provided, from switchboard to invoicing. Many companies have embarked on formal Total Quality Management(TQM) programmes under which they look at all quality and cost aspects for everything that is part of the supply chain for a service or a product. This critical assessment process is demanding new attitudes by everyone concerned to the quality of their work and to the approach that they are adopting in satisfying client or customer requirements. With this comes an awareness that *"Everyone is both a customer and a contractor, internally and externally"*.[8]

This is demanding that staff are very much closer to the client than has traditionally been the case in the past and ensures, in turn, that they have a better understanding of their client's needs and aspirations. This means devolving responsibilities to the staff concerned for the project as a whole;

---

[8] Third Age Careers - Meeting the Corporate Challenge. Curnow and McLean Fox. Gower, 1994

hence there is less need for tiered management structures with all their controls and authorisations. The result is a much slimmer organisation overall which is intended to be totally entrepreneurial and very close to the customer. This trend goes very much hand in hand with the general desire to downsize, and so flat management structures are now a common feature of many businesses today, with very few layers between the lowest staff level and the CEO.

Of that country which is renowned for its egalitarian approach(USA), in comparison with its European cousins with their stratified societies, I have heard it said that nobody needs to be more than four levels of contact away from the President. Many commercial organisations would be well advised to emulate this!

The trend will almost certainly continue as workers take on increased responsibility for the commercial as well as the technical aspects of businesses; clients are increasingly demanding a balanced approach between the technical and the cost(commercial) side of any service, so that the single-focus point of contact must have full responsibility for both aspects.

The introduction of a flatter management structure is usually part of a restructuring process; these have been going on with a vengeance in almost all UK business organisations since the 1980s, starting with many traditional manufacturing industries in the early '80s but spreading to the service sector in the middle/late part of the decade.*"Restructuring has an entrepreneurial thrust for companies trying to win in the global Olympics"*;[9] the sense of innovation that can be created by a well-managed restructuring process does create new opportunities and can emphasise the new attitude and relationship which companies are seeking to develop between their people and their clients.

The whole process can be seen as much more positive than a defence to market downturns, shedding assets and management layers in response to more competitive trading conditions; it is a real way of trying to find a new form for the organisation which will enable it to deliver more with less. *"Spending less but creating more value"* as Moss Kanter puts it.

These kinds of changes are something that all of us in business have had to learn to come to terms with, and indeed have become excited with them as they have opened up new ways of working. A state of perpetual change is definitely becoming the norm as businesses search for long-term survival mechanisms; these can only come about by being close, quick, efficient and

---

[9] When Giants Learn to Dance. R. Moss Kanter. Routledge, 1989

committed to a loyal client base. It can result in better work by fewer people; the people become more skilled but, more importantly, more committed.

All of this improves the cost effectiveness and hence the profitability of a company. But there is a right way of doing it, and the benefits only accrue to those who can adapt and thrive in this new kind of business climate; some recent research[10] illustrates the downside of this in several companies which had adopted this approach:-

- 53% of affected staff were dissatisfied after the reorganisation
- 60% of staff felt that their career development prospects had worsened
- there was a high turnover of staff, 35% over 2 years, from the sample companies
- good leadership, essential in this kind of change, was not apparent in 40% of the companies.

4.2.3. Localisation and regionalisation

The need to be much closer to the client than perhaps had been the case, say 15-20 years ago, has been the driving force behind the geographic spread of offices that is now characteristic of almost any consultancy of any size, including the specialists. In 1980, a large firm of Consulting Engineers might have had between 4 and 10 offices throughout the UK; now it would typically have 15-20, and although there has been an element of growth in size which accounts for part of this increase, even very much smaller firms have found it necessary to have a spread of regional operations so that they can continue to serve their established clientele locally wherever the client requires.

Exactly the same trend has occurred across much of manufacturing industry in the UK where there has been a need for the business to address and be close to a regionally based clientele. Perhaps one of the best examples of this is that of the new Toyota car plant that has been constructed at Burnaston, near Derby. Derby already has a significant engineering industry, based to a large extent on Rolls Royce and the former British Rail Technical Centre, but the introduction of a major automotive plant, with its emphasis on 'just in time' supply chain management, has spawned the rapid growth of new component suppliers in the area. Just the same phenomenon has been observed in the case of the new Citroen car plant recently erected in China; the interest amongst French component manufacturers in establishing new factories in its vicinity has been amazing to see.

---

[10] Roffey Park Management Institute, Newsletter No 4, Volume 9, Winter 1996/7

At the same time, there has always been the recognition that there are very real advantages in many industries in concentrating skills in a single location so as to maintain critical mass in particular disciplines and achieve an overall efficiency. This mitigates against the dispersal strategy demanded by clients.

These factors, also, have impacted on the management structure of service industries, with the need to ensure that regional managers have the autonomy that clients have come to expect without the need for decisions to be referred to the 'Head Office'. At the same time that this provides a high degree of autonomy for local managers, and, in some cases, full bottom-line profit responsibility, it has forced organisations to look seriously at how their overall management structure copes with this autonomy whilst at the same time continuing to ensure that there is significant 'value-added' from having a network of offices. These gradually become self-supporting and can easily tend to generalise and then overlap/duplicate with other group offices in specialist skill functions, creating internal competition unless carefully managed and motivated.

The new corporate ideal involves a smaller fixed core but requires an extended set of partnerlike ties; this all gives new *"scope for internal alliances, deal-making and leverage"*.[11] The trick is to ensure that all of these aspects work positively for the corporate whole. With this type of organisational streamlining, the central team then has to rely on the localised individual operating units to be its 'eyes and ears', i.e. to provide the market intelligence which it can interpret from the perspective of the company as a whole. This move also impacts on the pressure under which the local or field units exist, but the function can be used as an effective way of keeping the local offices in touch with the strategic thinking of the core.

4.2.4. Partnering and alliances
In decades gone by, it was common practice for companies to undertake just as much as they could of any project, be it in the manufacturing or service sector, so that the company in question was in control; it would manage the project in its entirety and would be responsible completely for the eventual outcome.

That has all changed in the last 30 years as projects have become progressively more complex and as management have accepted that their company might be good at doing some things but is unlikely to be good at

---

[11] When Giants Learn to Dance. R. Moss Kanter. Routledge, 1989

everything. It is now common-place for major industrial and manufacturing groups to link up with each other so as to share the costs (and the risks) of development work and enable the combination to have a more robust place in the market. Companies have learnt to become friendlier to each other, and, by pooling their resources, they have achieved more with less.

This trend has spread in recent years into the service sector, with a vengeance, so that it is now almost normal to find companies working together on complex or large projects or in complete fields of interest in a number of ways:-

- in cross-company consortia so as to undertake work that no single member could undertake on its own, with credibility,
- in joint venture, usually for a specific purpose and for a limited time
- in cross-ownerships by minority holdings to ensure the allegiance and the commitment of each party to the venture, and
- in strategic partnerships, perhaps with other members of the supply chain.

Even in the UK construction industry, with its traditional history of confrontation, claims, over-runs, etc., there is now a new spirit around with the publication of the Latham Report and its strong recommendations for a partnership approach to all construction projects.[12] The Private Finance Initiative (PFI) set up by the UK Government is also compelling very different groups of funders, constructors, operators, lawyers, Consulting Engineers, etc. to go into consortia together; this forces them to begin to have a common interest, separate from their own individual company priorities which are often in conflict.

All of these developments are demanding new attitudes amongst management who are, in the process, learning not only from their peers in other fields but also beginning to understand the key aspects of their colleagues' business. Everyone now needs friends in the global market; confrontation is definitely out!

---

[12] The Latham Report into the UK Construction Industry, Constructing the Team. HMSO, 1994

4.3. WORKING METHODS

4.3.1. The shamrock structure

So far, I have looked at some of the ways in which organisations are restructuring themselves by downsizing, creating flat structures with few intermediate layers of management, giving local autonomy and reorganising for bottom-line profit right through the company. All of this requires new approaches to working and new attitudes on the part of management.

Turning now to the ways in which people can be expected to work within these new structures, one of the more interesting descriptions is perhaps what Charles Handy calls the *'Shamrock Organisation'*.[13] The shamrock has 3 leaves, and hence its relevance to an organisational structure which Handy portrays, having:-

- a **central core** of highly paid, highly dedicated personnel, working probably in a pressurised environment; core staff are absolutely essential to the business.
- a **contracted fringe** of individuals and organisations that work on a contracted basis for the core. They are paid by results, not for time put in, and ideally work in a virtual continuous partnership with the core so as to maintain some continuity.
- **hired outsiders** who are brought in for peak workload. These tend not to be people off the labour market who are looking for work but people who have chosen to work on an individual assignment basis, perhaps part-time, and who may have a long-standing, although informal, relationship with the core.

This type of working arrangement poses a significant management challenge, much greater than having all staff in the company on the permanent payroll. Different techniques now become necessary to manage and motivate each component of the organisation. At the same time, it is a challenge to management to identify each component correctly and not permit a drift so that people fall into the wrong categories, either by chance, or, more easily, as a way of solving other short-term management issues.

4.3.2. Portfolio working and the Third Age

One of the trends in employment that has been very evident in recent years has been the tendency for employers to achieve staff reductions and restructuring of their businesses by simply disposing of the older members of

---

[13] The Age of Unreason. Charles Handy. Arrow, 1995

the workforce. This has been particularly obvious in the former nationalised industries which have now been privatised, and in the public sector, where the numbers of payroll employees in the over-50s bracket have become very small.

During recent conversations with a large outplacement agency for executives, mainly in the over-45s age bracket, I was told that whereas 10 years ago 95% of their clients had been looking for new full-time employment, now only around 70% were in this category. The remainder were looking for 'portfolio' work, i.e. a range of salaried or consultancy positions but all part-time; these could be as part-time executives, retained consultants or simply specialist advisors whose skills were only required by the employing company perhaps for 1 or 2 days a week or even less on a monthly basis. Such people would also take on freelance consultancy assignments on a project basis.

There are obvious attractions to this type of work, particularly for those in the 'Third Age' who are perhaps looking for a more varied career later in life. However, the other advantages of this type of work lie in the fact that it may well be more secure than a single full-time job since it is less likely that all the jobs would come to an end at the same time; additionally, this type of working makes perhaps better use of such people with their wealth of experience rather than for them to be employed simply for their probably narrower range of technical skills.

There are also advantages from the employer's point of view in these arrangements; they can continue to call on the experience of older members of staff as they require, rather than having to keep them on the full-time payroll when their experience might not be fully utilised.

In their book on the Third Age (op.cit.), Curnow and Fox have given examples of the best practice in this area as used by a (sadly) limited number of larger employers ranging from Guinness to Marks & Spencer. Of particular interest to professionals is probably IBM's 'Skills Rebalancing Offer', a programme under which older employees had the option of joining a new company, Skillbase Ltd., as independent part-time consultants for guaranteed amounts of work for a number of years after being asked to retire from IBM. Skillbase then uses and markets the expertise of these people, partly to sell back to IBM, sometimes even on a full-time basis, but also to sell elsewhere in the market; it has proved to be a most successful venture commercially, whilst IBM have preserved access to proven expertise when they require it.

This example serves as a warning to other companies, who in their drive to de-layer and reduce costs, have adopted a wholesale policy of 'releasing' all of the older generation. Whilst, in the short term, this is an attractive way of reducing head-count and salary costs, it:-

- often incurs substantial redundancy costs, thus blighting immediate performance figures, and
- loses hard-won expertise in perpetuity.

This latter aspect could be a concern because as economic activity picks up, and as the demographic time-bomb of reduced numbers of young people coming into the labour market gathers force; it will also only be a matter of time before some clients perceive that the 'depth' of experience, that intrinsic 'I know what it should look like' overview, has been lost. For the relatively conservative construction industry, this could be a significant factor and one which needs to be handled with care; the IBM type of approach minimises risk in this area.

A further advantage, at the personal level, of the IBM type of arrangement is that outsourcing in this way can give new status to the people concerned in the eyes of the original employer and of their former colleagues who might still be on the permanent payroll. Now, the new freelancers are respected as partners in, and contributors to, the business, rather than as expensive 'has beens' or 'hangers-on'. They often also earn more in this new type of arrangement; certainly they have more flexibility in their life.

All of these types of arrangements do mean that employers need to have flexible employment and management structures to cope with people wanting increasingly to work in this way as they get older; they may also have to adapt to possible conflicts of interest when portfolio workers have jobs for competitor employers in a similar field of work.

4.3.3. Teleworking

"Joe has gone home to finish that report in peace and quiet, without interruptions - he'll be in on Friday when it's finished ". That must be a phrase which all of us are familiar with, the classic escape from the distracting hustle and bustle of the office; and above all, the escape from the telephone that respects no fixed appointments or commitments.

As more and more people use home computers, as fax, voice and E-mail and the Internet become readily available on a home basis, so the prospect is opened up for many people to do much of their work from home with only occasional visits to an office as may be needed. This will apply more in some industries than others, and it has been common practice in marketing and sales functions for many years in some trades. For Consulting Engineers, the

example quoted by Handy in 'The Age of Unreason' about his friend Walter
might be particularly appropriate:-

*" Walter runs a design and consultancy business with a staff of around 100
professionals - quite big. He runs it from a converted warehouse, except that he
hasn't converted it very much. There are no offices in it. There are meeting rooms,
there is a superb farmhouse kitchen, there are drawing boards scattered around,
there are word-processors, telephones and computers abounding but no one, not
even Walter himself, has any private space - except for the secretaries, who are
really not secretaries as such but project co-ordinators.....*

*Walter told me, 'I don't want my designers and consultants spending their
time here in this very expensive space. I would rather they were out with the client
or working at home where I will provide any equipment they want. They only come
in here for meetings, to use some specialist equipment and, generally, to keep in
touch. We lay on the best breakfast in town in that kitchen of ours.....It's a working
club really'.*

*...It is the perfect facility for a network of individuals linked into a small
core..."*

One well-known computer service company, the FI Group plc, with a
turnover of almost £100m. in 1997 and a salaried payroll of over 1000, uses
'distributed office' techniques under which the major proportion, around 70%,
of their staff work at home or at locations that are convenient to them, rather
than in the offices of the main company. Large numbers of the staff also
work on a part-time or irregular basis, so that staff capacity and workload
are kept in constant balance.

By early in this century, a significant proportion of the working
population is likely to have these kinds of varied working patterns; people
will be seeking more independence in their work, and organisations will be
encouraging their employees to take firm control of their own careers, their
finances and their lifestyle. British Telecom reckoned that, in 1995, more
than 2m. people in the UK were already working from home for more than 3
days per week and that, by the 2000, this number would have risen to over
4m. - a significant proportion of the total workforce.[14]

4.3.4. The flexible office and the hot desk

Earlier in this chapter I have shown how more flexible ways of working are
becoming standard practice in many companies; these have implications for
the office facilities that need to be provided by any company for its
workforce.

---

[14] The Guardian, 2nd March 1996

In the service sector, overheads, of all kinds, might represent anything between 50% and 150% of gross payroll costs according to the industry and organisation involved; in turn, office and accommodation costs might account for up to a half of this overhead at the lower end of the range. These costs are therefore highly significant in a competitive environment.

In addition to the 'Walters' of this world, who want as many of their staff either with clients or working at home for clients, there are several other professions, the accountants and management consultants perhaps being the most obvious, where staff spend large amounts of time away from their own office, in meetings, on the client's premises, writing reports in a quiet environment, etc.

From these considerations and the new flexible working patterns that are emerging, the concept of 'hot desking' or the 'mobile' or 'flexible' office has devolved. In its simplest form, staff 'book' a desk in the office for each occasion when they work at their office location and simply move their personal effects to that location for the day in question. They check in with the telephone, logistic and information systems on their arrival. In the days when most people only require access to a terminal on the network, this achieves significant cost savings. It also ensures that visitors do not have that dispiriting experience of passing lots of empty desks along their route to their meeting - that always raises questions as to whether the absent staff really exist or whether something more sinister is afoot!

Before Brown & Root recently moved into new purpose-built offices for its engineering consultancy operations, it surveyed the usage of its office space and concluded that about 17% of its new space could be devoted to flexible working. Yet the PA Consulting Group go much further than this. They estimate that, in their consultancy, as many as half of all workers will 'hot desk', spending just half of their working time actually in the office.[15] Whilst engineering consultancy may not go quite as far as this, clearly there is a very large trend in this respect.

Another example of the flexible, yet fully equipped, office is one where access is unrestricted for staff and the office is fully functional 24hrs./day. This is sometimes particularly welcomed by younger people working together on projects where the excitement and fulfilment of the project give them an urge and incentive; they can work together on the project, with colleagues, at times that suit their other - social - activities. These kinds of measures can significantly increase productivity for a particular project.

---

[15] Management Consultancy, January 1997

At the other end of the spectrum is the fully-equipped mobile office. When asked if he was about to vacate his space in a busy car park, the driver said, *"No, I'll be here for another couple of hours"*. He had his portable computer, mobile fax and phone at the ready; he didn't need a fixed site for his work that day - or perhaps any day.

It is the amazing development of information technology that is responsible for many of the changes in practice that I have outlined here, and, no doubt, further steps in technological development are just around the corner for all of us. Without sophisticated communication and computer networks, many of these practices would simply not be feasible.

4.4. EMPLOYMENT ARRANGEMENTS
4.4.1. Staff ownership and employee participation

Alongside the increased role that workers in many companies now find themselves taking up, there is pressure for them to have a financial stake in the business. Ideally, this pressure comes from both directions, but where the existing owners are resistant to the trend, there is marked pressure from the workers. In the large plc's, this takes the form of share options, but in the more professional world of consultancy, there can be a demand for a much more meaningful involvement in what is, after all, a company dependent almost entirely on simply the 'knowledge' skills of the workers. They, of course, are free to work elsewhere at any time; when they leave, they take their particular skills, and perhaps their previous employer's reputation, plus even the clients on occasion, with them.

I have already quoted the example of the FI Group with their flexible approach to employment, but perhaps a more important aspect of this company is their ownership policy. The company was originally set up by Steve Shirley in the 1960s, but after two decades of rapid growth, she took the decision in the late 1980s that it was time to offer the staff ownership of the company. This was for 3 main reasons:-

- the **business** reason - *"doing better can only be achieved through our people. Our people must be given the freedom to apply their professional skills and business acumen in a way that keeps them a step ahead of our competitors"*. This means giving them a stake in the business.
- the **people** reason - the company worked on the basis of a high level of trust and confidence in others. Nobody wanted the company exposed to the vagaries of changes in leadership and style; staff ownership could control this.

- the **moral** reason - Steve Shirley again, *"in FI, people are important, important enough for them to have the utmost that I can offer - namely, ownership".*[16]

In this way, the company's co-operative culture has been enhanced by a co-ownership philosophy; the firm has now become a Plc with significant staff equity participation.

4.4.2. The Team Approach
A good leader can manage a team of 50 with ease, a poor leader a team of 1 with difficulty, it has been observed; the concept of teamwork is one that now pervades most industries, with teams, typically, being any size from 5 to 25. It is felt that, above this number, the personal dynamic of the leader begins to be lost; spontaneity, enthusiasm and innovation become replaced by rule following, fixed job cultures and routine procedures and systems.

Hierarchical Structure	Team-based Structure
Strong leader	Shared leadership roles
Individual accountability	Individual & mutual accountability
Individual work products	Collective work products
Leader runs meetings	Open discussions for all team members; all are active in problem solving
Often conflicting approaches to completing work. Cost and value are often vague; delivery is dictated	Common, agreed-on approach to completing work. All agree critical delivery points and costs

Table 4.1 *Comparison of Team Structures*

---

[16] The Journey to Empowerment. Mrs. Steve Shirley. Paper at IPA Annual Conference, 1993

Within teams, all staff need to, using that horrible jargon word, be 'empowered'. This enables them to be free to innovate, to be creative and responsive to clients, whilst still *within* the management culture of the company. Many modern management theorists extol the virtues of this team and empowering culture and perceive that this is the way to structure the company of the future. In this culture, the role of the managers changes to that of being team leaders; each manager becomes a leader, coach and example for his/her team. Table 4.1 illustrates *some* of the contrasting attitudes which apply in comparing team-led organisations to those that are more conventionally structured.

4.4.3. Employment contracts and incentivisation
Much as company structures are changing, so also are employment conditions to reflect the more flexible working practices which are perceived to be of benefit to both employer and employee.

For that core of full-time employees on the payroll, it is still very much a traditional employment contract of salary, perks (where relevant), and company pension scheme - usually, but not always, of the contributory type. Recent changes in UK taxation are reducing the incentive effects of perks such as company cars, whilst the ability to now get some personal tax relief on personal pension schemes, whilst in employment, is decreasing the attraction of the company scheme; this then becomes less of a useful way of first attracting, and then motivating and retaining, staff. The changes have all been subtle and gradual over the last 10 years.

I have already discussed the place of share ownership in one's employing company as a motivator. An alternative approach has been to use profit-related share schemes; to date most of these seem to me to be fairly notional in terms of remuneration quantum, other than at Board level, but they could be extended to improve the incentive effect. In the UK, their tax benefits become added incentives only for employees in companies that are operating at a reasonable profit level and that are confident of improving their profitability on a regular year-on-year basis; that poses difficulties for much of the construction industry whose profitability is notoriously variable.

At the same time, more companies, particularly in the financial services sector, are moving their staff on to a combination of salary (generally adequate) plus significant bonuses based on the year's trading results. Again this is intended as a motivator, and a strong one at that, but it forces a concentration on short-term performance and it is questionable whether this policy sits happily alongside the quest for total client satisfaction over the longer term. There is much experimentation taking place across most industries in this area.

A final thought under the heading of incentives is the concept of 'team' rather than purely 'individual' pay. In a recent survey carried out by the Industrial Society, only 10% of the 366 training managers surveyed worked in firms that offered any form of team pay. But 52% of those surveyed believed that team-based pay was the best way of promoting team building; the same percentage of employees believed that as well![17]

4.4.4. Training and career development

Until recently, the general notion of training and career development was that this was a subject that employer and employee sat down to discuss at perhaps annual intervals, with the employer very much taking the lead as to what additional skills he/she perceived would benefit the company which the employee could then go on to acquire, usually at the employer's expense. This approach is still common in many companies.

However, there is now the recognition that no longer is it the employer's sole responsibility, although it may well be in the company's interest, to ensure that staff are adequately training themselves with continuous improvement and learning. Within the more recent individualistic culture of our society, this aspect of a person's skill is definitely becoming an individual responsibility. *"So just forget about 'finishing' your education. Defend your career by developing a better package of knowledge and skills than the next person"*.[18] With the need to train, and retrain, perhaps several times during the course of a lifetime's career, it is clearly in the employee's interest to get on and do it him or her self. The role of the employer should still be supportive, but, to a large extent, it is the employee nowadays who should be taking the initiative.

Thus employees should no longer find their security from being employed; they find it by ensuring that they are fully employable. This attitude will lead to a new self-confidence and a clear direction for the individual of:-

- where they want to go,
- where they see that their company is going, and
- are they content and confident with both?

---

[17] The Guardian, 2nd March 1996

[18] New Work Habits for a radically changing World. Price Pritchett. Pritchett & Associates, Inc., Dallas, Texas, 1994

This way of thinking should give a new energy to individuals within a company and a new mood of confidence generally.

4.5. CHANGE PROGRAMMES

Lots of companies, including many leading firms of professionals, have embarked on large internal change programmes during recent years and several aspects of these have been illustrated in this chapter. These have not been so much about business processes as about trying to change behaviour and behavioural attitudes. They have been designed to make a much better job, than the traditional structures, of harnessing the talents of individuals, so that they can contribute in broader and richer ways to the company as a whole.

These processes have usually had the same basic steps in their overall approach:-

- **Step 1** - Get acceptance of the overall plan of what is proposed and its objectives - usually no easy matter!
- **Step 2** - Set up a number of initiatives, such as:-
    - internal communications as to progress, etc.
    - put in some new I.T. so as provide resources for people
    - focus on client care
    - install widespread business planning
    - set up personal development and appraisal systems
    - sales skills training for many staff
    - monitor internal morale, by surveys
    - have plenty of internal workshops for team-building.
- **Step 3** - Monitor the programme and adjust things as it develops.

These kinds of programme seem to have two main effects on staff. Either people start to work things out for themselves, i.e. they construct the route map, then the vision evolves and the company can move forward. Alternatively, there is an overall cynicism about the management's 'new idea' and a general reluctance to look at anything new; if this becomes the case, then more radical changes need to be imposed to bring people to the appropriate starting position..

4.6. SUMMARY

These are some of the changes that are now evident in the way in which businesses are beginning to organise themselves and in which they are going to continue to change. Consulting Engineers are not going to be exempt from these pressures in their own profession, and indeed a few companies are already adopting some of these new ways. However, the pace of change is so rapid that many will have to speed up their rate of change; those who have hardly started will have to indeed move quickly.

What is becoming more and more evident, throughout this whole debate, is that companies cannot *own* people's intelligence; that belongs to the individuals concerned, and they need to be persuaded and encouraged to use it, in a continuing and more dedicated way, on behalf of the employing organisation. Failure to provide the appropriate employment framework will not provide that encouragement and they will go or drift elsewhere, taking not only their skills, but probably also their clients with them.

It is a combination of this new freedom of key employees, coupled with the Information Technology Revolution and the more open, demanding and competitive markets throughout the world, that is the catalyst to continuous and significant change in all service industries.

And, increasingly, it is the case in knowledge organisations, like engineering consultancies, that staff will almost always know more than the managers; they will have the client contact and many of the other business connections. Managing effectively in such a situation is quite an art.

It is not going to be easy to incorporate all of these new ways into service organisations. Drucker, in his book 'Managing for the Future',[19] talks of the service sector with its knowledge/service workers as being the real key for future national prosperity within the developed countries. He shows that there are many areas for real productivity improvements which need to be grasped if the nation in question is to remain an important economic force for the next century; more importantly, he shows that this improvement has to be ongoing, year on year - there can be no standing still! This is now even more of a challenge for managers!

I now start to look at how Consulting Engineers can grapple with these aspects in their own particular profession and thus chart a way forward for themselves.

---

[19] Managing for the Future. Peter F. Drucker. Butterworth-Heinemann, 1992

NOTES ON THE CHAPTER

CHAPTER 5

Market requirements of the future

In this chapter I look briefly at the likely trends in future markets for Consulting Engineers. Although what follows is primarily from a UK perspective, many of these trends are of a global nature, so that they have a relevance to Consulting Engineers from many different countries. These trends all have implications for their organisations if they are to develop to meet the markets which are expected in terms of size and complexity, client requirements for skills and attitudes, and the commercial response which must follow from all of these.

5.1. FUTURE MARKETS
5.1.1. UK domestic market
Chapter 2 has highlighted many of the recent trends in the UK market and it seems likely that these trends will continue. There will be a slowing down of the rate at which public sector in-house organisations are transferred to the private sector as this process reaches completion or as a result of possible changes in government policy, but the same trend in the private sector will continue; thus there will be continued disturbance in the competitive framework of the market, with new firms and competitors being created or major firms making acquisitions. At the same time, there are likely to be continued constraints on public expenditure and further growth of new ways of undertaking projects utilising private finance. In the private sector, growth will be dependent on the general state of the economy but a modest and

gradual improvement could be in prospect; again there will be a search for new ways of undertaking projects, affecting the way in which risk is carried by consultants.

The lesson which we have all learnt from the experiences of the late 1980's and the 1990's is that few clients are any longer content to simply entrust projects to a Consulting Engineer or, indeed, to any professional, without clear and committed ideas as to project scope, time and cost. This has come about from a general perception that too many projects end up as too elaborate or too simplistic, over-budget, or delayed in implementation; in other words, they are simply not fit for the purpose for which they were originally conceived. The credibility of the profession as a whole, and indeed much of the construction industry, has been seriously damaged in this area; this has led to the rising influence of specialist project management companies, alongside totally new ways of undertaking projects, so as to achieve the project objectives which the client has always wanted.

The availability of finance, or restrictions on this in the public sector particularly, has been another force for change; this has led to the development of the Private Finance Initiative (PFI) in the UK and similar forms of contract elsewhere whereby clients can distance themselves from much of the construction or operational risk within a project. It is then those who are involved in project execution, including the Consulting Engineers, who are asked to carry a share of these newer risks; this has implications for consultants' financial strength and performance.

Another area of potentially significant growth - although it has been heralded for some time but has yet to really take off - is in projects that have a strong environmental component, whether this be in the form of new projects or improvements needed to reduce and overcome existing environmental degradation. There seems to be little doubt that the 'green agenda' is rising in importance with its consequential expenditure.

This has been particularly evident in the water supply and sewerage sector, where privatisation of the industry has been accompanied by demands by the Regulator, OFWAT, as to new standards that are to be achieved by the newly privatised companies and the consequent amount of investment that they have to make. Landfill and waste disposal is a further area which has attracted a lot of attention, with recycling becoming a 'buzz' word in many local districts.

Even the developer sector has not been immune from these pressures, with the need for them to start to take into account such concepts as 'sustainable development' and recycling of 'brown' land in urban areas, quite apart from the long-established principle of planning gain, which often takes the form of environmental improvements.

Industry, too, is having to improve its pollution controls to meet new environmental standards and to take pollution hazards seriously into account in its operational risk assessments.

Environmental concerns are also going to have a major impact on the transport sector where there is now strong pressure to achieve meaningful improvements in both urban and inter-urban transport systems without colossal capital investment, with its normal corollary - significant environmental disturbance.

All types of projects are demanding an increasingly wide range of skills and the involvement of highly technical input and innovation, often more in the software, control and process elements of the project rather than simply in major new construction.

All of these aspects are suggestive of the need for significant change within the whole of the construction industry, and in consultancy practices generally, as they tackle the domestic market, and much of this type of work and thinking is already evident in some of the more progressive and larger consultancies. These factors are also leading to the establishment of a whole host of small consultancies which seek to specialise in one or other of these new areas of work, particularly on the software and control side.

The UK market is, and will remain, large. The total UK construction market is around £55bn. per year, and that this is likely to be reasonably static over the next few years; of this, some £32bn. is new work, with the balance being more of a repair and maintenance nature. Most new work requires significant consultancy involvement one way or another, whereas other forms of work require less involvement, so that there might be an engineering consultancy turnover within this overall market of perhaps £3bn. per year.[1] However, this market, although very large by almost any standard, is very over-crowded; it is this factor which is driving the ferocious competition which is so characteristic of more conventional domestic consultancy. These pressures are likely to remain, and the only way in which the excess capacity can be dissipated will be from reduced numbers of individuals in the profession, particularly engineers; this could take a generation to work through, unless there are significant external sources of work outside the UK, and I now turn to these.

---

[1] Note: The Construction Industry Council estimated that the turnover of **all** professional services in construction for the UK was around £6.7bn. in 1994.

5.1.2. European market

The Atkins report for the EU (1993) foresaw that in the late 1990s there would be a steady upward trend in construction output within the EU, although more recent trends would seem to indicate that this has not been the case, as public finances, in particular, in the larger member states have tightened with the advent of monetary union. However, there are now signs that these restrictions are beginning to ease, as Euroland emerges from recession, and construction is now expected to grow quite firmly over the next few years, particularly in the civil engineering sector.[2] Primarily, these effects are, or will be felt, most noticeably by consultants in their own domestic markets.

So far as cross-border work is concerned:-

- there has been a relatively small amount of cross-border work within the EU in the consultancy field so far, in spite of the Services Directives. This appears to be due to an inherent lack of open-ness in client bodies, cultural and language differences, etc; it seems unlikely that this situation will change rapidly, and so the opportunities for this type of work will not increase quickly, in spite of the pressure which the EU continues to put on member states to really open their markets in a meaningful way to foreign consultants.

- with the recent and further proposed enlargement of the EU, there will be more competitive pressures in the market, particularly from Scandinavian consultants used to operating in English. This will tend to work to the disadvantage of UK or other English-speaking consultants because of their traditional weakness in language skills.

However, there seems to be an expectation of significantly increased opportunities in Central Europe as the local economies stabilise and as new investment comes through. These countries, however, do have good technical skills already to a large extent, so that it will be in the economic, safety, environmental and project management disciplines where the best opportunities will lie. This will favour the very large or more specialist consultancies mainly.

Given current trends in Russia and the CIS, it seems likely that any increased market for outside firms will be slow to develop other than in the funded sector or for major industrial development projects. Again, most of these will require a wide spread of disciplines.

---

[2] Forecast from Euroconstruct as reported in NCE, January 13th, 2000.

The appropriate approach for consultants for work in Europe as a whole is likely to be on the basis of collaborative effort with other locally-based or specialist firms, with many of the changes demanded by the UK market being replicated on the European scene. An additional and important component will be that of appropriate language skills. The European scene will demand that work should be done in the local language, other than perhaps on large aid-funded projects. This is a real area of weakness for UK and American firms in comparison with most of their European competitors, so that some form of partnering will be essential.

Many features of the UK market are therefore relevant in an European context, but the language component and a willingness to co-operate with other competitors will be distinctive features. This will require a change in culture on the part of many businesses, in whatever field they work, but the change will need to be particularly dramatic amongst engineering consultants for whom these are, to a large extent, very different and new ways of working. Those firms which have the ability to change and which have widespread language skills should be able to increase their work within Europe steadily.

5.1.3. The world market

Over the last 10 years, the trend in global economic terms has been for the newly industrialised countries, NIC's, of Asia in particular, to grow at the expense of European, North American and Japanese economies, and this is likely to continue over the next 15 - 20 years. The effect of this is that, for example, the UK share of the world economy is likely to fall from 4.2% in 1994 to 3.2% in 2010, whilst that of Germany will fall from 8.4% to 5.9% over the same time frame; even the USA share will fall from 27.4% to 21.1%. On the other hand, China's share will increase from its current level of 2.1% to 8.2% and Taiwan's economy in 2010 could be equal to that of the UK.[3]

However, the domestic markets for engineering consultancy in developed countries seem likely to remain reasonably stable, with many of the changes that are prevalent within the UK becoming the norm elsewhere. Thus, Consulting Engineers will face the same challenges in terms of different patterns of working.

Within the NIC's and the developing world, the engineering consultancy market has already become quite well-developed, in many cases, for

---

[3] Article in Management Today, April 1996, by David Smith

indigenous firms of Consulting Engineers and so the days of the expatriate consultancy must be numbered, with the exception, possibly, of more management-orientated skills.

In the face of this shift in economic activity, new ways of working would seem to be imperative if UK consultancies, or indeed consultancies from any major country, are to retain anything like their current share of global activity outside their home territory. There are two distinct features in the changes in the consultancy market that have begun to take place:-

- **Firstly**, consultants have recognised that they can obtain a lower cost base by utilising local skills and that these skills are increasingly available in many overseas countries; a number of firms are using this approach, not just for world projects but for domestic projects as well. The effect is to take the routine design work out of the home territory of the company concerned and to develop networking/distance-working skills within Consulting Engineering as well as adapting to cultural differences in project development. This trend is likely to continue.

- **Secondly**, for aid-funded projects in the developing world - once the preserve of UK consultants in the English-speaking world and of their European competitors in their own previous colonial territories, e.g. the Dutch in Indonesia and the French in sub-Saharan Africa - there is increasing competition as the register of approved consultants with the World Bank grows by the addition of consultants from newer countries. Aid funds are increasingly addressed towards social, training and maintenance issues rather than to new infrastructure projects, and these require a wider range of skills, many of which are new to established consultants. Aid funds are subject to more price competition and to budget reductions and the local component of any project is often now very significant in project award.

At the same time, there is a reluctance on the part of recipient countries to use foreign nationals whose qualifications do not demonstrably exceed those of their own people. The importance of the c.v. of each outside expert has become critical to obtaining and undertaking these projects, particularly the length of experience in the type of work and in the region in question. This factor is preventing traditional consultants from providing training and experience to the younger members of their staff, particularly in the more traditional engineering skills which are now available to a high standard in many developing countries. The age profile of 'approved' experts is therefore gradually lengthening.

These factors all present a gloomy picture for consultants unless they change the whole basis of their involvement in this type of work; this needs now to be based much more on collaboration and co-operation with local consultants, even to the extent of encouraging the local entity to take the lead in projects with only specialist input from the foreign firm. There has to be an acceptance, still resisted by some, that the opportunity to bring work back to the home-base, i.e. direct exports, is going to become very limited. This thinking is providing the pressure for the development of global consultancies.

Overall, the role of foreign engineering consultants in these parts of the world will become increasingly advisory rather than technically orientated.

The same comment is likely to apply, in effect, to all of the more mature markets of the world, e.g. USA/Canada, Europe and Australasia, where the opportunity for any foreign consultant to make significant penetration into well-developed local markets must be limited. However, again, given the right approach to partnering and appropriate specialist technology, there will be one-off opportunities for involvement.

5.2. THE QUALITY AND REGULATORY FRAMEWORK

Until around 15 years ago, it was sufficient for most reputable firms of Consulting Engineers to have a core of properly qualified staff, a rigorous checking system and sufficient computer aids for design to ensure that the company produced an adequate product on a fairly consistent basis. In the last 10 years, in particular, all this has changed, as clients have increasingly demanded:-

- better and more formalised quality control systems,
- reduced fees for routine design work,
- demonstrable value for money on the total project, and
- shorter time frames for both the design and implementation stages of a project.

At the same time, UK and EU regulations have been produced in large numbers in an attempt to improve safety standards on sites and in offices, whilst at the same time ensuring more consistency of approach across the wider market of the EU.

5.2.1. Quality assurance

Formal accredited quality systems have been common-place in manufacturing industry for many years, but it was not until the mid-1980s that UK firms of Consulting Engineers began to take seriously the need for rigorous Quality Assurance procedures under the then BS 5750 and, subsequently, the ISO 9000 series.

There has been debate in the construction industry as to whether the Standard, originally apparently conceived for the manufacturing sector, could really be applicable to construction, particularly to design activities. In fact, however, ISO 9000 *"does little more than describe 20 elements of good business practice, and identify areas where problems are likely to occur"*[4] and so, for many organisations, conforming to the Standard is simply a matter of formalising existing good practices. Nevertheless, the debate continues, although many large clients are increasingly requiring the full adoption of the Standard by all of their suppliers as a precondition to being on approved tender lists.

It's my experience that the adoption of these quality systems has been something of a mixed blessing:-

- in the first place, for the smaller or less well-organised larger companies, the introduction of Q.A. has indeed led to more rigorous procedures which have improved the quality of their work and, in turn, given them greater credibility in the market.

- on the other hand, for the larger and often better organised companies, the Q.A. system has not really improved their existing checking and management control systems, although it has introduced a greater element of standardisation and formal documentation. The display of the Q.A. logo adds to credibility, but for the larger firms, this was hardly in doubt anyhow. It has, undoubtedly in many instances, led to a proliferation of paperwork, due to the need for a better defined audit trail under the Standard's procedures.

Some research[5] carried out a few years ago on the introduction of Q.A. schemes in the UK showed that:-

- 89% of firms reported improved operational efficiency
- 76% of firms improved their marketing effectiveness.

However, this was a broad sample of companies, and I would suspect that there were few engineering consultancies, if any, within it; the improved efficiency figures for Consulting Engineers are, in fact, probably very much

---

[4] M. Whetton in Civil Engineering Proceedings, Institution of Civil Engineers, 1996

[5] A survey of quality consultancy schemes. Pera International and Salford University, 1992

lower, although the improvement in marketing effectiveness is a very significant benefit which many consultants might wish to achieve.

Q.A., or some later variant, is, however, here to stay, and firms would be well advised, now that it is common-place in the whole of industry, to re-examine their procedures to see whether these can be simplified, thus saving in routine documentation but without negating the effects of the procedure. After the initial excitement of the introduction of Q.A., now is probably a good time to re-examine its workings in each individual firm to see whether procedures can be simplified without losing the intended result of the standard. Companies need to be aware of the 'stultifying' effect that rigid adherence to ISO 9000 procedures might have, because the procedures themselves tend to inhibit change and hence, perhaps, creativity.

5.2.2. Safety

As in the field of quality, where it is the well-organised firms that have benefitted probably least from the introduction of a Q.A. system, so, in the field of safety, it is the smaller or less well-organised companies which have benefitted most from the introduction of new safety standards and the CDM Regulations[6] for construction processes. Again, this is all a part of the attempt to improve on-site safety and site working procedures generally, and the requirement to now have a Planning Supervisor for every project has, in many instances, led to additional involvement, and sometimes additional fee, on the part of the consultant, although additional training for this role should also be required.

The proper certification of both office and site procedures under the Health and Safety Regulations has undoubtedly improved safety generally; this general move has led to a real growth in work in the safety and risk assessment areas.

5.2.3. Total Quality Management(TQM) -or Value Engineering

Coverage of potential new ways of working for consultancy practices would not be complete without a look at the way in which this subject has affected the way in which projects are handled. Originally a concept only applicable to the oil, gas, and petrochemical industries, this often now comprises a total review of a project so as to satisfy both the consultant and the client that the optimum scheme is being proposed in terms of:-

- value for money

---

[6] Construction (Design and Management) Regulations. HMSO, 1994

- fitness for purpose
- optimum whole-life costings
- reduced risk of error
- assured project delivery
- certainty of project outcome, including subsequent operations.

With the growing pressure on all forms of resource in every sector of the market, it's clear to me that all companies will, in future, have to comply with the majority of these elements for almost all of their projects; some already do this, but those who don't will have to.

TQM was originally mainly concerned with achieving the 'right product for the right cost at the right time'. It is now, however, seen as part of a much wider movement to change the total culture within organisations so that it brings continuous and real improvements to everything that the company does. It starts to tackle the causes of problems rather than solve them once they have occurred; it aims to 'get things right, first time, every time', thus improving both internal efficiency and client/customer satisfaction at the same time.

TQM is thus something which affects and involves the whole organisation. Its adoption as a method of working has significant training implications for everyone; it starts to ensure that the customer, and everything that the company does for him/her, really does come first.

5.3. COMMERCIAL REQUIREMENTS FOR NEW PROJECTS
5.3.1. General trends
This chapter has already identified several areas, particularly finance, where Consulting Engineers are being asked to take a share in project risk. This is partly to demonstrate in a tangible way their commitment to the project and belief in the integrity of their own work, but also partly as an element within the process of offloading risk from the ultimate client. The days when a client would accept the costs of all 'unforeseen circumstances' in a project are long gone; now clients consider, often with good reason, that the Consulting Engineer, or certainly the consultancy and construction team as a whole, are better placed to assess and allow for such risks. Whether, as in some instances, the construction team can really be expected to allow for 'war risk' is a moot point, but that may simply be an over-reaction by clients to taking on any risk at all, and a corresponding measure of the desperation of the construction industry for turnover, almost at any price - or risk!

At the same time, such a commitment to the 'downside' of a project has to be balanced by the potential 'upside'; team members need to have the opportunity to share in project success when this occurs.

Shared commitment is also a better way of achieving 'bonding' and co-operation between the various members of a project team, in that there is a shared interest in achieving success. In many a 'failed' project during the 1980s, particularly in the building sector, one of the frequent factors in the failure was the inability of the various parties to the project to co-operate and resolve difficulties together in the interests of the project as a whole; all too often, each took a defensive and unco-operative stand in order to try and protect their own liability. Clients have tired of picking up the costs, often quite enormous, and other effects, resulting from internal squabbles within diverse project teams.

In recognition of these factors, and the general adversarial nature of work in the construction industry, the UK Government announced, in 1993, a joint review of the construction industry and the subsequent Latham Report - "Constructing the Team," HMSO, 1994, - produced new guidelines for co-operative working. Progress on its implementation has been disappointingly slow; the construction industry remains fairly adversarial in its approach to business.

Following on from this, within the UK, there has been a lot of discussion and experiment in new ways of tackling projects, so as to:-

- avoid over-runs in cost - endemic for most major projects in the UK
- avoid slippage to programmes, with subsequent financial consequences
- keep the client and the implementing team 'at peace' with each other.

Two new ways of undertaking projects have emerged in earnest, the Private Finance Initiative, PFI, and Partnering. Both of these methods have, as one of their main objectives, the whole of the project team working together and being related to one another in several aspects.

5.3.2. PFI Projects
In the PFI concept, a promoting group will include:-
- project analysts
- designers
- project managers
- constructors and operators
- financiers and investment appraisers
- legal and contractual advisers.

The precise form of contract has varied from DBOM(Design, build, operate and maintain) to DBFO(Design, build, finance and operate) and BOOT(Build, operate, own and transfer), but, in essence, the principles of the project and the range of disciplines needed have been similar. In all these variants, engineers can have key roles in forecasting usage of the new facility, if it is a major piece of infrastructure, and in designing the facility 'fit for purpose'. They may also be involved in the appraisal of construction and maintenance costs and methods.

Some of these projects have now been completed within the UK and are now well into their operational phase, which can last for anything from 25 to 75 years, according to the terms of the concession. Some trends are beginning to emerge as to how successful these arrangements are for the participating companies, who have often got together at an early stage so as to first prequalify and then bid for the project.

It seems that the groups are often quite difficult to hold together, mainly because each participant does have a different interest in the outcome of his element of the project, and either groupings are beginning to emerge which have an interest in only a certain stage of the job or companies are beginning to establish themselves which can undertake the whole project in its entirety, backed up by appropriate financing. Both of these developments are tending to diminish the conventional role of Consulting Engineers in the project - many of their services can simply be bought in on an ad-hoc basis, particularly the design skills, by the protaganists in the project team; civil engineers need to be aware of the danger of losing any 'lead role'.

However, for those consultants willing to adapt and change, there is most definitely an increased opportunity for work in assessing feasibility, whole-life costings and usage; many of these tasks require sophisticated modelling and synthesising skills. Only once the project has been fully defined and secured does the engineering design commence; even after this, there can be an ongoing role with maintenance and usage surveys and further improvement strategies.

With this type of project, the Consulting Engineer generally works within the total group promoting the project, and some consultants have made it a definite policy objective to develop a portfolio of PFI projects, thus providing an ongoing workload over the length of the concession; however, there is also the opportunity to work in a different role and to use their technical skills to re-assure other, perhaps outside, investors in the project that all is well. Thus, Consulting Engineers can add a significant level of comfort in that what has been done is satisfactory and sound. This kind of approach requires a broadening of the traditional client base and a wider understanding by staff members of how other professions, for example, financiers, perceive, quantify, and deal with risk.

5.3.3. Partnering projects

The Partnering approach is not totally new to the UK and has been in use in the USA, particularly in the petro-chemical field, for several years. Here it is the oil industries themselves which took the lead in forcing participants in a project to co-operate more effectively together, under the threat of project cancellation if they did not achieve the savings which the companies believed could result from an integrated approach to project development and implementation. The results of this initiative have been startling, with cost savings of up to 45%, on the Andrews Field Project, having been achieved in comparison with original traditional project estimates; there have also been significant time savings. And all participants in the process, not just the commissioning client, have ended up being happy with the outcome! Whether there has been some kind of 'honeymoon' effect, with savings on later projects being much more modest, only time will tell.

This kind of approach is picked up in the Egan Report, "Rethinking Construction",[7] which is a follow-up to the earlier Latham Report. This concludes that a move away from competitive tendering towards partnering and alliances is likely to lead to a significant improvement in the way that projects are completed. This must be true for the project that 'goes well', but it is a more testing time for the partners when the project runs into difficulties, as is so often the case in civil engineering. There are also the siren voices that warn against the development of 'cosy' cartels within a partnering arrangement such that it is the client who misses out. Clearly, it is desirable for there always to be some element of competition somewhere in the procurement process so as to ensure that the best value for money is being obtained.

The concept of partnering is so new, for most of the industry, that there is considerable scepticism as to its long-term benefits and practicality. There are now a number of demonstration projects being undertaken in the UK that aim to establish standards, procedures and benchmarks for working in this way.[8] These seem to suggest that there are indeed savings to be made with this approach, albeit not on the scale of the Andrews Field job. Yet other

---

[7] The Egan Report, Rethinking Construction. DETR/HMSO, 1998
[8] Movement for Innovation @ www.m4i.org.uk

voices suggest, more cynically, that this is yet a further way of clients eroding their suppliers, i.e. contractors and consultants, margins.

However, most people seem to agree that some way must be found to reduce the traditional antagonistic approach within construction which must be counter-productive and wasteful of resources in the longer term. The attempt to change this suggests that the guiding principles for the future must be to work together or stay stuck in an isolated role where there is not only a highly competitive field but also one where there is limited opportunity to contribute added value to the project as a whole. Bedelian[9] summarises this emphasis on partnering in the wider construction industry as a shift from *"a 'revenue enhancing' culture to a 'productivity' culture"* where all parties in the relationship concentrate more on producing what the client wants at the price which he can afford, rather than in maximising their own particular contribution. Consulting Engineers will not be able to escape this shift!

All of these changes require new attitudes and new approaches throughout the construction industry, both within companies and between them, for partnering, with all its potential, is neither a panacea nor an easy option. There is a forecast that, in the future, for every £6m. of project turnover in the UK, there will be one or more alliances within the total project team.[10] This will involve new ways of looking at potential partners, and, especially in the public sector with its demand for open-ness and accountability, a framework for establishing relationships within a project. Such a framework is shown in Fig. 5.1.

The diagram indicates the (fairly) extensive work needed to set up the process for an individual project and how, in the public sector at least, the decision about, and the form of, partnering has to follow the project team selection process, i.e. there needs to be some element of pricing fixed at an early stage, before the partnering arrangements are finalised in place.

There seems little doubt that the UK construction industry as a whole is in for a big change, as all project participants, including the Consulting Engineers, begin to realise that they are going to have to have a stake in the project, with all the benefits as well as the liabilities which flow from that involvement. These pressures, quite apart from innovative forms of contract like the PFI, will compel those Consulting Engineers who wish to participate in major projects to find ways of creating stronger balance sheets and capital structures so that they can take on the necessary financial risks. Given

---

[9] Proceedings of the Institution of Civil Engineers, August 1996

[10] Address by Jordan Lewis, reported in New Civil Engineer, 20th February 1997

current margins in traditional consulting engineering work, it would appear that very few companies are going to be in a position to take on such responsibilities with their present financial structures.

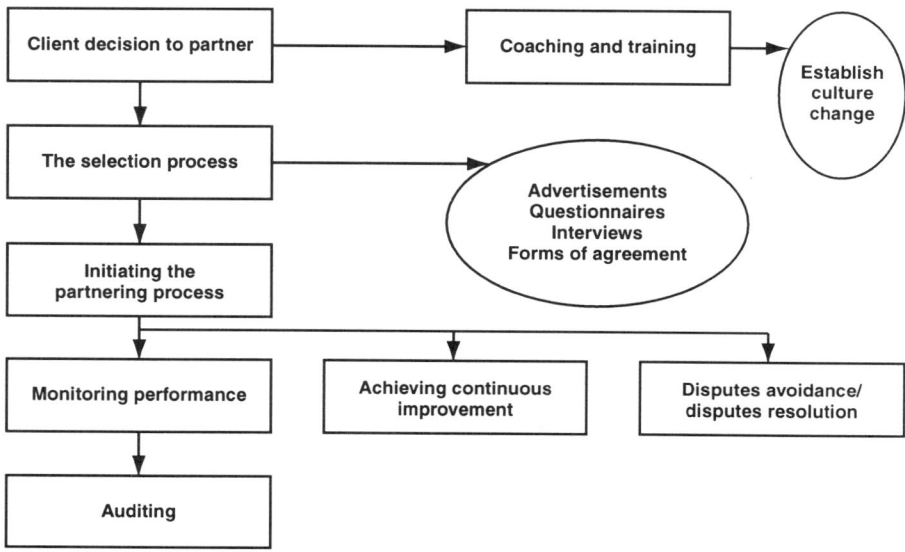

Figure 5.1 *The Partnering Process in the Public Sector*[11]

(Source - ICE)

This all leads to the conclusion that some further wholesale restructuring of the industry is in sight, with probably outside sources of finance, and hence control and influence, being brought into the profession as never before.

A parallel trend is also emerging, certainly within the UK market and, I suspect, within many other countries in Europe and the USA, whereby large and influential clients are beginning to want to have semi-permanent partnering arrangements with a strictly limited number of consultants. Such arrangements demand heavy investment in marketing, etc. on the part of the consultant, and in selection processes on the client's side; the intention is to acheive mechanisms for regular contracts for a whole variety of activity.

To some extent, this type of process can fly in the face of a requirement for more open competition and only time will tell whether it serves the client

---

[11] Partnering in the Public Sector, European Construction Institute, 1997

well or not. What however, it does lead to, is to force smaller/medium-sized consultants to team up on a regular basis so as to gain a place on the prized short-list of consultants chosen by the client.

5.4. THE PROJECT LEADER/MANAGER

As team structures have grown in complexity, so the client has become increasingly reluctant, in many cases, to take responsibility for the interface between the different disciplines, and this has led to the emergence of the single point of contact or focus within a professional team so far as the client is concerned. Consulting Engineers have often perceived that it's in their interest to take on this role, if only to safeguard their position as the lead engineer for the project. They are also usually quite good at it!

By way of example, when such a lead role is not held on to, there is the instance of major building projects where, until perhaps 15 years ago, the lead professional role was almost invariably taken by the architect. This function has declined very significantly over the intervening period because architects have not been able to demonstrate their lead role and provide the financial and resource control for the project; this role has been taken over almost entirely by project managers and quantity surveyors, or by engineers, but to a lesser extent. It seems that engineers also were a little slow to recognise the trend and its implication for their own relationship with the client. A number of engineering consultancies are now beginning to try to increase their involvement in this role and are seeking to recapture it; this gives them the direct contact with the client which is really essential for the full understanding of a project and of the client's aspirations for it. Above all, it enables firms to 'manage' the client relationship rather than simply be at the behest of some other company that is working more directly for the client.

The same trend has been detected in the USA, as illustrated in an article by Gary Kellogg. He comments that the market-place wants engineering leadership, but he detects a trend for engineers to become increasingly specialised and to abandon their wider responsibilities to the profession and to society as a whole. He comments:-

"Civil engineers bemoan the lack of respect and yet at the same time do little to take responsibility for their actions. Others, not engineers, then step into the construction market-place as leaders. Civil engineers do not have leadership because they do not take leadership"[12]

---

[12] Proceedings of the Institution of Civil Engineers, August 1996

5.5. RANGE OF SKILLS & TECHNOLOGY

All of this suggests the need for an increased range of skills in engineering consultancies and some very different approaches to project work, both within and outside the domestic market. Abroad, the opportunities will continue to lie mainly with the larger multi-discipline or very specialist consultancies, whereas in the UK and other home markets, the differing market patterns which we've seen over the last few years can be expected very much to continue.

This impacts on the particular ranges of skills that it is appropriate for Consulting Engineers to carry, on the way in which they co-operate with each other and on the way in which they organise themselves internally so as to be lean and responsive to the new shape of the market. These latter factors are particularly relevant to the organisational structures of Consulting Engineers who have always had to maintain fairly flexible approaches to staff employment because of the propensity of Government in the UK to use the construction industry as a significant regulator in the national economy.

Since the late 1960s, in the days of the Roskill Commission for the Third London Airport, it has been recognised that infrastructure projects produce benefits and disbenefits, some of which can be costed, e.g. the valuation of journey-time savings to the traveller, and some of which are intangible - the famous example of the calculated disbenefit resulting from the demolition of a Norman church on the line of the project, springs to mind.

During the 1980s it became clear that for works funded from the public purse, more serious attention had to be paid to the socio-economic impacts of projects, including such items as employment and planning effects, environmental impacts and competitive effects. This approach now also applies to all large or significant development projects, whether in the public or private sector, and of virtually whatever type; this process is driving project development work to become totally multi-disciplinary, with the engineering skills being relevant, alongside other equally important disciplines, at the feasibility stage but not becoming the dominant skill until the implementation phase of the project. Further examples of the kinds of activities which consultants might get involved in and which all add intellectual value to a project are:-

- sustainability issues
- modelling and synthesis
- whole-life costings
- funding mechanisms
- all forms of 'soft' technology
- operational and maintenance issues and techniques

The range of skills that are commonly required for almost any important project, and indeed for many smaller projects that impact on the community in any way, could therefore include:-

- project managers
- planners, both strategic and at project level
- engineers, almost all disciplines
- economists, for demand forecasts, financial and economic appraisal
- operators and maintainers of the facility
- environmental experts, e.g. landscape, wildlife, pollution, global effects
- construction implementation experts and cost managers
- safety and risk assessment experts
- financial advisers
- sociologists
- personnel/human resources advisers and trainers
- public relations advisors.

This list is by no means comprehensive, but the assembly and retention of just some of these skills pose real logistical burdens on firms if they wish to have in-house the range of disciplines demanded by many projects. Of key importance, in their role of project preparation, are the project managers - experts who are often drawn from an engineering background originally.

Consulting Engineers will have to decide which of these skills they should carry in-house and which can be acquired on a project basis without the risk of loss of quality or project cohesion; alternatively, they need to decide which can be acquired by establishing one-off joint-ventures or by establishing longer-term strategic partnerships with appropriate specialists or competitors.

It's noticeable that a variety of approaches are being adopted to tackle these issues:-

- development of practices by an ever-widening base of in-house skills,
- growth of regular partnerships and strategic alliances between consultants so that the combined team can provide all of the necessary skills for a project, and
- conversion from being pure consultancy organisations to groups which aim to manage the total provision of infrastructure; this is a development similar to the way in which much of the oil and gas industry grew and now traditionally works.

At the same time, many of the less engineering-orientated disciplines are being carried by other leading professional firms, clearly intent on exploiting new project opportunities and seeking a total project involvement; this particularly applies to the leading firms of accountants and management consultants, many of whom are now beginning to have a number of engineers on their staffs. In this way, they are acquiring sufficient engineering expertise in-house to meet the immediate and up-front demands of a project, whilst the detailed engineering, to just take one example, is let off on a (highly) competitive basis to the more traditional consultancies; these firms are thus removed from direct contact with the client and are often left out from the initial and conceptual phases of projects.

In considering the range of skills that are needed within any engineering consultancy organisation, thought has also to be given to the speed at which routine design work has often to be carried out. A recent report quoted:-

"It will be possible to design a project in weeks rather than months through the use of knowledge-based engineering and object-based modelling".[13]

This same report identified how it is only through leading-edge technology and involvement in novel and more effective forms of project processes that UK consultancies are going to hold on to their present share in the global market in the face of intense local and other international competitors. Five main sectors were highlighted as offering real opportunities for development and growth:-

- coastal and river engineering and management
- environmental improvement and sustainable development
- infrastructure for urban development and megacities
- transport planning and infrastructure
- water and waste-water engineering.

The comment was made, however, that this latter sector would require UK firms to move their technology a real step forward, perhaps in the face of enhanced skills that are already apparent within other international competitors. However, where UK firms did seem to possess an advantage was in their experience of resolving conflicts between demand and the protection of the environment; this apparently gives them a very credible starting point.

---

[13] Thriving in a Global Market: Technology Strategies for Civil Engineering Exports. The Institution of Civil Engineers, 1998

5.6. OVERALL SUMMARY OF MARKET REQUIREMENTS

The outcome of all of these changes can be summarised in strategic terms:-

-consultants will have to be prepared to take on part of the project risk, probably in addition to their fee component. This will affect their financial(balance sheet) structure and will encourage new forms of partnering to share the risk on larger projects or to handle the prolonged timescales involved with DBOM/concession type projects. Unwillingness to take on risk that is outside a consultant's core discipline may, conversely, lead to a narrowing of specialisms in firms.

- for industrial and other private sector projects, it will increasingly be the project-management skills that will be the main driving force in the client's procurement strategy, rather than pure technical ability. Again, this will impact on the importance of this discipline in comparison with more technical skills within firms.

- there will be continued disturbance and uncertainty in the market as a whole as a result of competitive pressures and restructuring, particularly in the public sector. This will enhance the pressure on, and the need for, dedicated marketing policies within firms so as to maintain track of relevant client movements and penetrate new organisations effectively.

- continual response and adjustment to shifts in the pattern of domestic markets as totally-open competition perhaps gives way to competition based more on partnering suitability, credibility and consultancy profile.

- for many projects, clients will increasingly want a single interface approach. Large consultancies will have to have a broad range of skills to satisfy this need; they will have to address how to acquire new skills and how to retain and manage them, once acquired.

- the increasing emphasis on environmental safeguards, etc. will lead to a very much wider range of disciplines being involved in new projects than has been the case in the past.

- many of the new disciplines which Consulting Engineers will need to carry, or associate with, for their projects will be software, control, or I.T. based rather than just traditional design and construction skills.

- there will be many opportunities for niche players in the market, offering very specialist skills, particularly on the software and I.T. sides; they will often have to present themselves in linked ventures in order to appeal to clients direct or act as sub-consultants to the larger firms.

- so as to hold on to their present share in the global market, consultancies will have to keep their technology ahead of the general field.

- smaller firms will concentrate on their more local base and on smaller projects/clients. Again there will be a demand for management skills from them if they are to lead projects, and a need to form linkages with others in order to offer the total range of skills required. This could lead to more widespread practices whereby the larger firms take on a project in its entirety but subcontract all of the routine technical work to a local firm.

- all consultancies will have to comply with Directives issued by both the EU and the UK Government in recent years. This will include all safety, products and workplace directives. Again, new skills may need to be developed and incorporated as standard routines within commissions in addition to procedures such as quality assurance, risk assessment and safety case statements, which are already well-established.

- above all, engineering consultancies will need to demonstrate and implement a willingness to co-operate with other firms of consultants and even with their competitors so as to achieve a more sophisticated skill-mix and the financial strength which will be essential to successful involvement in projects in a 'lead' or client-related role. For companies who adopt these philosophies, it will be very much a time of change, reflecting:-
- new attitudes amongst technical and managerial staff
- new roles and forms of project involvement
- acceptance of different risks
- new relationships with other companies and with different forms of clients
- greater financial involvement in projects.

These types of trend will be the foundation for new strategies for Consulting Engineers. Most of them will necessitate significant change within companies and a large increase in staff training and technology development so that firms become prime leaders in the new ways of working.

NOTES ON THE CHAPTER

CHAPTER 6

Strategies for Engineering Consultants

6.1. STRATEGIC MANAGEMENT
6.1.1. Overview
The preparation of business strategies is always an ongoing process and an iterative one; it is not an activity that can be done once, then filed away. There are far too many companies throughout industry where the company's strategy is either in the owner's head (a somewhat risky location!) or filed in the MD's confidential filing cabinet - this means that the rest of the organisation soldiers on in ignorance of the 'grand design', whilst the strategy itself gets progressively more out of date and superfluous to setting the day-to-day direction and objectives of the business. Strategies prepared or considered in this way are often treated as one-off exercises, tend to be somewhat academic and usually miss the essential element of strategy implementation and continuous review.

In their book on corporate strategy[1], Johnson and Scholes set out a basic model for the strategic management process and I reproduce this in Fig. 6.1 overleaf. This demonstrates the iterative and ongoing nature of the whole process of strategy formulation for a business; Consulting Engineers are going to have to adopt models like this for their own strategic management processes if they are to make the changes outlined in Chapter 5 and continue to change and adapt to a fast-moving market and industry.

---

[1] Exploring Corporate Strategy. Johnson and Scholes. Prentice Hall, 1993

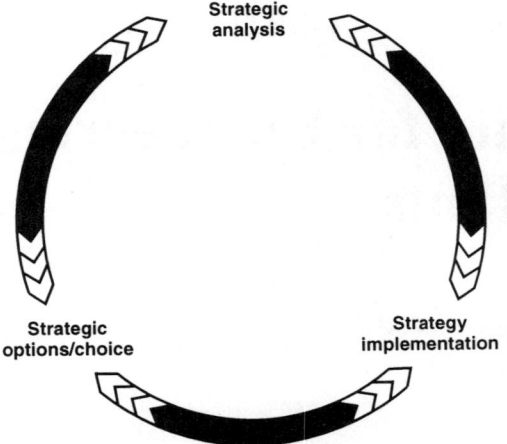

Figure 6.1 *A basic model of the strategic management process*
(after Johnson & Scholes)

Obviously, each firm of Consulting Engineers will derive a different strategy for itself and it is not the purpose of this book to produce the ideal blueprint for the business. That will follow, logically, from the position and the opportunities which any one firm sees for itself in the market, through to how it organises itself, both internally and externally, to address that market. The whole process has to be iterative. However, all will need to follow the same basic process in their strategy formulation so as to derive an individual strategy which will mark them out in the market.

Fig. 6.2, again from Johnson and Scholes' work, illustrates the elements in a strategic management process in greater detail.

6.1.2. Strategic analysis
This looks at the strategic position of a company within its market in terms of the opportunities and threats in the market, the company's resources, its internal strengths and weaknesses in terms of staff numbers, experience and skills, and the expectations of various stakeholders; these could include parent companies, senior managers, funders, owners and, preferably, staff generally.

It is unlikely that there will be a complete match between the current direction of a business and the picture which begins to emerge from this strategic analysis; it is the extent to which there is any mismatch that confronts the management of all companies with the dilemmas of change.

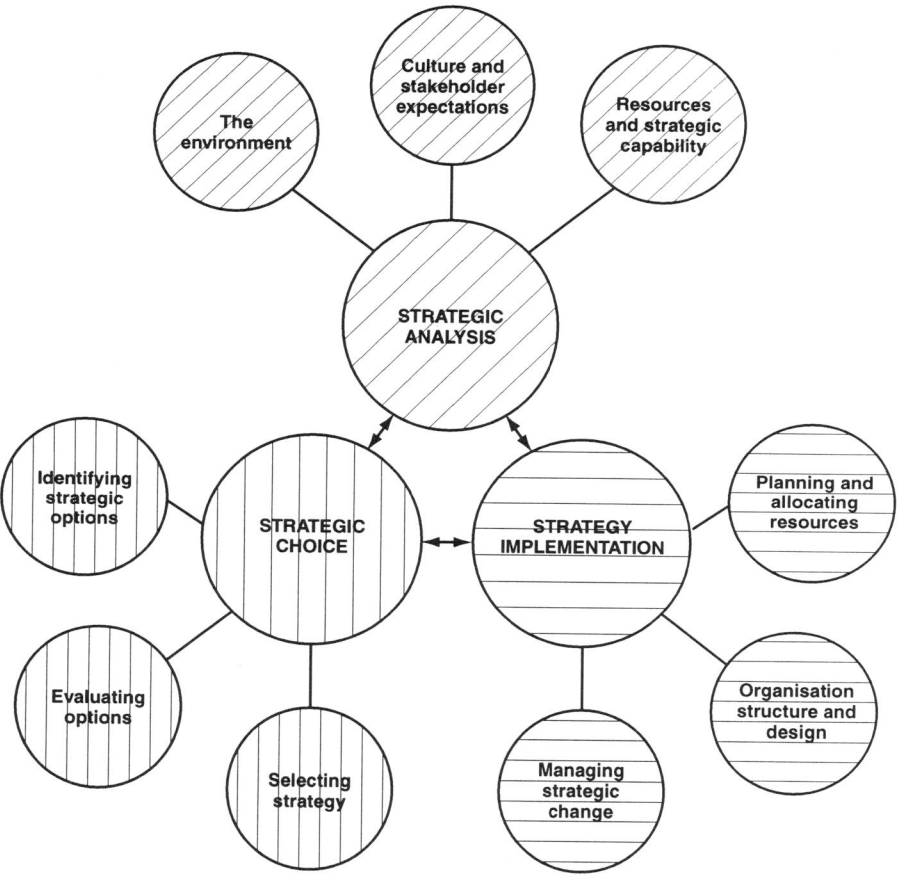

Figure 6.2 *Elements in strategic management*

(after Johnson & Scholes)

6.1.3. Strategic choices/options

It's at this stage that the process moves on to considering strategic choices, by first generating options, evaluating them and then selecting preferences. There are unlikely to be clear winners/losers in these options because there will be some element of risk and uncertainty in almost whatever strategic option is chosen, so that the choice is likely to be, in the end, a matter of management judgement. To this extent, final choices will therefore be strongly influenced by the values of the managers, and of other influential groups within the company, as well as by the cultural heritage which the business might carry within itself. Final choices may thus well reflect the power structure within the organisation and its tradition.

6.1.4. Strategic implementation

This phase is concerned with devising ways of implementing the strategic options which have been identified and selected by the company - in essence, how will the company have to change so as to be able to implement the new strategy? I consider that one of the key areas, for professional firms like Consulting Engineers, will be the extent to which changes in the organisational structure are needed to carry through the strategy.

As I said earlier, this whole process has to be iterative, with the consequences which flow from changes in organisational structure and the adoption of specific strategic options being tested again at the analysis stage against the resources and capabilities which are actually available within the company or which can be readily acquired.

However, a word of caution about sophisticated, and probably therefore time-consuming, strategy formulation:- *"Business situations and opportunities simply change too quickly for there to be much point in loading ourselves with piles of strategic speculation"*[2] - Rupert Murdoch. As always, there has to be a balance between brevity and comprehensiveness in strategy formulation.

In this chapter, I look at the key components of the Analysis phase, whilst in Chapter 7 I look at the critical element of organisational and management structure

6.2. SCOPE OF ACTIVITIES AND THE MARKET CLIMATE
6.2.1. First steps in the strategy

In the last chapter, I set out the general trends in the market for engineering consultancy, illustrated many of the likely market requirements for the future, and suggested that consultancies will need to look at all of these; there are certainly many options for them to consider. This process is the first step in the production of a strategy for any business, for it is the market and the demand for an organisation's products which is the starting point of any plan for the future. The response of all companies to these trends will differ, depending on where they themselves are coming from, but the general sequence will need to be:-

- Assess the market, first for general trends and then for detailed trends in the company's area of interest
- Assess the company's strengths and weaknesses in the market

---

[2] A speech by Rupert Murdoch to the International Institute of Communications, Sept. 1988

- Assess the opportunities and threats in the market
- Draw up preferences for the way in which the consultancy wants to operate in terms of disciplines, project services, employment, methods of working, etc.

Much of this takes the form of the classical SWOT analysis.

I discussed the importance and effects on national economies of improvements in the efficiencies of knowledge and service workers earlier in the book; the engineering consultancy industry is a clear example where such improvements have had a significant effect on performance, in terms both of each individual company, and, in the light of the high public sector content of much engineering consultancy, of the UK as a whole. The competitive environment which has been introduced into the UK market during the last decade has forced consultants to become very much more efficient and cost-effective, and this can be seen in the level of project fee levels where the same, or often an enhanced, service has been provided at dramatically reduced cost over the period; fee levels for comparable project services are now running at anything between a half and a third of earlier levels in many instances.

Whether the same **quality** of service is currently provided may well now be a moot point, and certainly there would appear to be a number of clients who are recognising that there is a very definite linkage between fee level and service quality; the public sector, in particular, has difficulty in adjudicating between such trade-offs, and it is in this sector that the most dramatic fee reductions have generally been seen.

In spite of this possible caveat, however, there is little doubt that the industry as a whole has become very much more cost-effective in recent years and has thus made some form of contribution to the overall efficiency improvement in service industries that Drucker emphasised. His theme was that such improvements need to be continuous in the present global competitive environment, and so consultants will be forced to address these kinds of issues on an ongoing basis. Potential fee levels and profit margins for different activities are therefore going to be important criteria in the selection of the most desirable future activities for a consultancy.

6.2.2. New services/new clients
As an example of this, some consultancies have recognised that there are significant business opportunities to be grasped and, in their view, good profit potential, in many service elements within the public sector, particularly within the organisations of Local Government in the UK. The

move to diversification and the opportunity that such work might provide by way of downstream follow-on activities, thus enhancing the turnover and capability of the mainstream engineering consultancy, has been one of the motivating forces behind this trend. However, other consultancies have taken this a significant step further and have moved into becoming serious providers of facilities management services across a whole range of businesses and industries; they perceive potentially large profit and business opportunities in this type of work and have concluded that significantly enhanced margins can be made by providing services such as building, office and estate management services, even the provision of virtually complete sectors of local government administrative machinery in some instances.

Whilst there is no doubt that good money can be made in this area at the moment, it does have a number of management implications for the consultancy concerned. For example:-

- this type of activity is very different from that in which most firms have established a track record and so they are no longer able to draw on their own past expertise,
- there is likely to be a somewhat different client base to that with which any consultancy has normally been used to dealing,
- the nature of the work, with its emphasis on repetition, maintenance and cost efficiency, is very different from the more one-off project-based activities of traditional firms,
- and, finally, the nature of many of the staff may well be more akin to 'blue-collar' organisations rather than the very definite 'white collar' of consultancy firms; Drucker makes the significant distinction between 'service' and 'knowledge' staff here and draws comparisons with industrial production processes.

All of this suggests that these companies are moving, not only into new disciplines, but also into very new markets, something that I will come on to in a moment. It is also questionable, in my view, as to whether such a strategy will significantly bolster existing traditional engineering or technically-orientated activities, as there would seem to be relatively little coherence between the two areas of work; an ability to develop a very efficient portfolio of general office services or facilities management would seem, to me, to have little connection with high-technology consultancy. Certainly such strategic moves will require some very different management and personnel skills within any acquiring consultancy, and it may well be that firms may have to be making distinct choices in these areas if they are to have a focused approach to their business. This is likely to become a significant issue in the Strategic Implementation stage of the process, when change implementation strategies are being developed.

A further issue in looking at these new kinds of activities is that, because they are new, there are few yardsticks or expectations about costs amongst client bodies. As these activities become more mature, costs and profitability can be expected to fall, year-on-year, in similar ways to those in which engineering consultancy fees have generally fallen.

These are the kind of questions which need to be looked at in connection with any move into new markets or services. My own experience of consultancy, and one that I have seen repeated over and over again, is that consultants often 'drift' into new disciplines which at first sight seem promising and which can sometimes be added on to other project work and thus *projected* as an additional lever for winning particular types of work. I am not convinced that the move into new disciplines or activities is always part of a definite and calculated long-term strategy; of course, as in any business, there must be the chance to take up good opportunities that are offered unexpectedly, and some of the most successful businesses of our time have been founded exactly on this entrepreneurial principle.

For those of us who may not have quite that 'streak of brilliance' which is the mark of the highly successful entrepreneur, a more methodical approach is probably required!

A typical diagram which illustrates a more methodical outcome to the selection of appropriate market sectors and services is shown in Fig. 6.3.

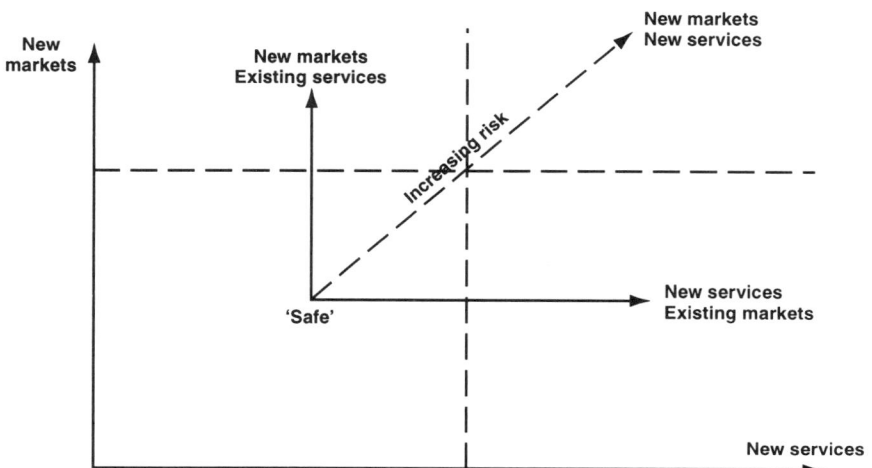

Figure 6.3 *The risks of changes in markets and services*

94

The diagram indicates that as a company moves its business simultaneously into both new markets and new services, it is taking on increased risk. Although this might, at first sight, seem to be an obvious comment, it is one which has often appeared to be ignored in many industries, with sad results. When such a shift in direction takes place, a firm no longer has the benefit of any history or continuity on which to base its experience and it has few client contacts; thus it exposes itself to considerably increased risk and costs, both in acquiring projects/commissions and in carrying them out. The analogy of moving one step at a time, i.e. remaining always firmly anchored on one leg, is another way of showing how a company can expand either its client base(one leg) **or** its range of services(the other leg), but not both at the same time; thus, risk is minimised.

When Touche Ross took a long and hard look at the way its business was organised and at the services which its clients required, it found that 90% of its revenues came from 10% of its client base[3]; this is a variant of the old 80/20 rule that for years has appeared to apply to many completely different aspects of civil engineering. For most engineering consultants, I suspect that similar findings would apply, thus suggesting that it is far more effective to focus on the services and the method of delivery of those services for the 10/20% of the client base when determining future business strategy, rather than tackling these questions across the whole range of the client base.

One of the characteristics of 'long-lived companies' is that, although they change, often quite significantly over time, they generally change in an incremental way, one step at a time, rather than in dramatic leaps and bounds. The evidence is that this is the way to maintain continuity, not only of the business, but of the staff and company culture as well.[4] And it appears to produce, generally, long-term profitability - hence survival!

Many firms of UK consulting engineers have had relatively long lives thus far, particularly some of the well-known names which have histories going back for 75-100 years. In contrast, nearly half of the top 50 UK companies in 1965 no longer exist and Dun & Bradstreet report that the median age of US firms is about 12 years.[5] So perhaps there is a word of warning here for those long-lived consultants who are considering very radical change in the scope of their activities!

---

[3] Management Consultancy, June 1997
[4] Seminar address by Arie de Geus at Roffey Park Management Institute, June 1996
[5] '100 Years and Counting'. S. Costa. American Management Association, December 1994

For all products and services, it is appropriate to carry out a full SWOT analysis. This results in an initial list of the best areas of activity for the firm based on its own experience and excellence relative to the market and to the opportunities that are perceived to be available, both in volume and profit terms, for the short/medium term and for the longer term.

6.2.3. Selection of activities

Another element in considering the scope of activities which a consulting company might provide is to decide whether, on a technical level, it wishes to undertake all project stages and whether it will carry out all of its work in-house or make use of subconsultants for elements of the project work. We are beginning, amongst UK consultancies, to get some different approaches to these questions and, as a result, very different cost structures, since both overheads and management structures vary according to the project stage for which a company provides services, in terms of:-

- feasibility and concept work, advisory work
- outline designs
- detailed engineering designs
- project management
- construction supervision
- facilities and operations management.

Table 6.4 is a simplified illustration of how a number of significant business factors might be assessed. At first glance, this particular illustration would suggest that the most promising areas of business are at the two extremes, the feasibility/advisory stage, requiring highly specialised technical input, and therefore very specialist staff, or the facilities/operations management stage, where there is less competition and a relatively low level of technical requirement. This type of table needs to be developed for different product lines or disciplines if it is to be used effectively as input to the strategy process. A full analysis would also include factors relating to the ease of entry to the market; this will be related to the market's perception of the company concerned - something which might be somewhat different to the company's own perception of its position in the market! Any exercise like this requires rigorous 'soul-searching' and a ruthless ability to set aside preconceived notions; it requires the courage to be able to sometimes say, *'we're not sure that we can do this particular thing well'*; then it's best to either leave that for others to do, or, perhaps preferably, line up others to do it on your behalf, thus still keeping a foot in the client's door.

FACTOR IN THE MARKET

ACTIVITIES	Ease of access	Profitability	Size of market	Competition	Level of technology and specialisation	Continuity potential	Management input	Risk
Feasibility, concept and advisory work	LOW	GOOD	SMALL	MEDIUM	HIGH	GOOD	SMALL	SMALL
Outline design work	MEDIUM	GOOD	SMALL	LARGE	MEDIUM	GOOD	MEDIUM	MEDIUM
Detailed design	GOOD	POOR	LARGE	LARGE	MEDIUM	MEDIUM	HIGH	HIGH
Project Management	LOW	FAIR	MEDIUM	MEDIUM	SMALL	SMALL	MEDIUM	HIGH
Construction supervision	MEDIUM	POOR	LARGE	LARGE	SMALL	SMALL	SMALL	SMALL
Facility & operations management	LOW	MEDIUM	MEDIUM	SMALL	SMALL	GOOD	MEDIUM	MEDIUM

Table 6.4   Assessment of Business Factors

KEY   favourable    unfavourable

The outcome of all of this exercise is a 'preferred' list of target markets and clients vs. a range of services which can be defined either on a project stage or on a discipline basis. The next step is to see which of these might be viable targets, given the starting point of the company and of its people. Alternatively, if the starting point of the consultancy is evidently far behind a particular market requirement, then there is a need to consider whether some form of strategic alliance or partnering could provide an appropriate entry position.

Finally, a key element in the exercise is to assess the relative profitability of the different activities; although there is no fixed rule, it is generally held that activities with a higher level of technology are likely to generate a higher margin of profit. However, there will always be the occasional market segment that has not been spotted by others, and which, although of relatively low technology, can give a high profit margin for the short time before competitors also adopt it.

In considering which activities to undertake, it needs to be borne in mind that clients usually look at the company's overall track record as well as its expertise in the relevant field. This suggests, for a large consultancy at any rate, that it needs to maintain a broad base of activities so as to maintain its overall credibility; it also needs to ensure, for example, that it maintains a sufficient range of traditional and mainstream engineering activities - sometimes called 'back-end engineering' - to provide credibility for its highly technical analysis, modelling and process work. Yet it is the latter which is likely to generate the profit and the reputation for innovation, on which the company's future must hang.

6.3. THE SKILLS OF THE PEOPLE

In their book on corporate strategy, Johnson and Scholes have set out 7 key areas which are associated with strategy and strategic decisions. Two of the most important of these relate to the 'scope of an organisation's activities' and the 'matching of the organisation's activities to its resource capability'. For consulting engineering companies, their resource capability is linked to the quality and skills of their people who are very much their prime asset base, and so this capability, together with the opportunities and requirements of the market, must be a further initial starting point for defining a firm's strategy for the future.

Thus, the next step is to assess the ability of the people of the company - its only real resource - either in terms of existing personnel or in terms of personnel who can be recruited/added to the organisation in order to match the demands of the services which appear to be most attractive.

Such work is personal, and the evaluation needs to be impartial and objective. It involves assessing all staff in terms of their key skills and abilities, something which probably few firms of Consulting Engineers have undertaken in a systematic and logical way. It is, however, the starting point for coming to an objective view of a company's potential, based on its current resources.

For the purpose of the exercise, skills can be categorised into 'behavioural' as well as 'technical' aspects, with technical skills being further subdivided on a technology base. The fields of assessment could include:-

- Technical skills - abilities, competence, specialisms - all the things which make up the 'good' engineer. This field also includes other professional skills, such as financial and legal abilities.
- Client skills - the ability to communicate, listen and respond to clients - an ability to manage the client relationship - typically project managers.
- Business skills - commercial awareness, foresight and ability to assess commercial/financial risks and opportunities. Market-focused people, again with financial and legal acumen.
- Co-operation skills - the ability to see opportunities for other people and to work with them so as to enhance the end-product. How can we get added value? The visionary element in a consultancy.
- Management/leadership skills - traditional management skills of overseeing, encouraging, empowering, leading and building teams.

Few people have all of these attributes in depth, and, in practice, few need all of them. But everyone within an organisation must have some, and preferably several, on a general level, with one or two in depth. Engineers are notorious for their inability to communicate effectively and there is no doubt, in my mind, that this has caused no end of damage to the profession within the UK. Some recent ACE research by Fisher Associates found, *"they, consulting engineers, need to match their high technical ability with an understanding of client needs, **which at present is low.**"*[6] The comment could hardly be more damning for a service industry! Equally, there is little point in putting a brilliant, but commercially unfocused, engineer into a position where he/she has significant management responsibility.

---

[6] John Bowcock writing in the Consulting Engineer, October 1995

Assessing the staff potential of any consultancy involves striking a balance between all these aspects, as well as being able to make estimates of people's ability to change and develop. On a technical level, various technical skills also need to be balanced against the market requirements for the projects which the company wishes to pursue.

Fig. 6.5 illustrates the likely, but hypothetical, picture of the importance of these various behavioural skills within a large UK engineering consultancy. The horizontal axis of the diagram reflects both the proportion of people in the various categories within the consultancy and the importance of that category of skill; these factors have has been combined within a weighted index. The derivation of the index for each particular firm is the task of management, alongside the importance of each behavioural skill.

What the diagram does indicate is the extreme dominance of technical, mainly engineering, skills in the present situation - a scenario which many UK consultancies are grappling with at present - whereas the requirement for future markets is such that it is the other skills which are likely to become almost equally dominant for the success of the consultancy and relevant to its ability to develop and change. Consultancies which are growing steadily may maintain the same 'absolute' numbers of engineering staff, for example, but it is likely to be the other disciplines which will fuel the growth.

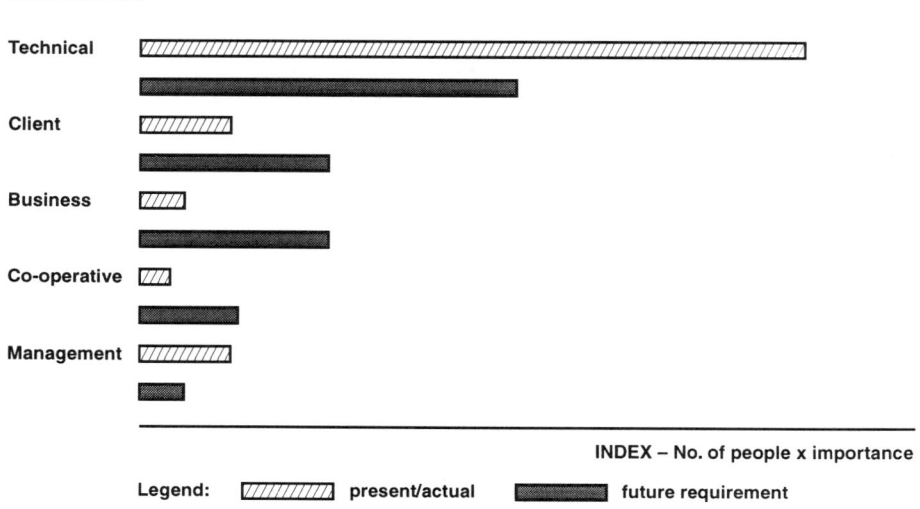

Figure 6.5 *Skill Matrix within a hypothetical engineering consultancy*

100

It is very much the potential ability of any firm to change which will dictate its ability to develop or even survive within its market; the process of change is one of continually adapting to emerging and changing markets by ensuring that, at all times, the correct mix of technical and behavioural skills is present within the company. Technical competence will continue to be a main ingredient in the success of a firm of engineering consultants, but business/marketing skills, visionary attitudes, and a flair for leadership and inspiring innovation, rather than simply a management ability to control and direct, will become crucial characteristics of the consultancy of the future.

The potential to change the mix of these skills and, indeed, the balance of all of the technical and professional specialisms, will very much control the ability of the company to adapt to the market environment. The push to change and diversify is strong; one of the largest UK consultants, W.S.Atkins plc., recently announced a 3-year plan to diversify its UK business, hand in hand with a 4-fold growth in overseas business.[7] One of the more significant comments, within this announcement, was that whilst the size of the company was expected to grow, the number of civil engineers would not change.

Similarly, within the big UK-based accountancy firms, there has been a trend to introduce more information technology and therefore reduce the number of pure accountancy staff whilst at the same time adopting a policy of recruiting more specialists in the fields that clients are wanting - including project engineers and project managers! This has had the effect of reducing the size of their graduate programmes, since graduates are not specialists; it has also led to the growth of employment of specialist freelancers who provide their services to the company, or indeed to a number of competing companies, as and when their skills are needed for the work that the companies have secured.

6.4. COMPANY CULTURE AND ATTITUDES

Management literature and articles on how other professionals are coping with changes in their markets, and, for the larger firms, globalisation issues, indicate that the work culture within a company is a crucial element in the future development and success of a company; earlier sections of this book have also illustrated some of the changes which are taking place which affect company culture.

---

[7] News. New Civil Engineer, 18th March 1999

In his book, "Managing for the Future"[8], Drucker spells out the need in the changing organisation to build confidence, loyalty and trust amongst all company staff, and he then goes on to emphasise how this is a very difficult task in today's changing world.

He uses the example of the way in which, in almost any company, the implementation of new I.T.systems has affected it. The evidence is that, as I.T. has taken hold in organisations and enabled them to down-size and have flatter structures, so the fall-out of middle-management jobs has caused alarm, despondency and depression at these levels. More jobs are created at the data-handling/technical end but a lot fewer in management as such. The middle-manager of the 1980s thus has few places to go, and only the very bright ones have managed to ascend the fiercely competitive upwards ladder; otherwise it leaves only some form of sideways move, perhaps into smaller companies, perhaps into self-employment - risky for lots of people, but often especially risky for these people - and portfolio working. The end result is that the company loses the loyalty and commitment of what was a key part of its work-force, it loses some of its sense of history and continuity, whilst the people concerned feel let down and rejected.

Similarly, there is an impact on these same people when the companies that they work for are involved in mergers or take-overs. So often, nowadays, it appears to be the 'deal-makers' and the 'merger-drivers' who are the main beneficiaries of take-overs, and it is the middle-managers, who have perhaps kept a company going through difficult times, who are made redundant when the new combined firm is set up. Nobody doubts that there is always a need for an element of reorganisation in these situations, but the long-standing client loyalty factor and company history are often neglected. History seems to be littered with examples of take-overs made in the interest of acquiring a new skill base or client portfolio; often, that skill base, client portfolio and relationship are all but invisible a few years later with the key people gone.

Yet there are companies who have been able to retain the culture of loyalty and commitment whilst still making these changes. How have they done it? I think that there are 5 main areas to look at:-

- Corporate work culture
- Attitudes of staff
- Participation and rewards
- Working practices and stress
- Attitudes to change.

---

[8] Managing for the Future, Peter Drucker. Butterworth Heinemann, 1992

Any successful strategy needs to be based on all of these factors, so it's important that they are addressed at an early stage in the process, as well as kept under continual review.

6.4.1. Corporate work culture

Work culture is a blend of ingredients, the basic ones being entrepreneurial spirit, incentives and motivation, team cohesion, and freedom, with responsibility.

All of these are human qualities which require a human, humane, caring and motivating organisation. Companies which achieve this seem to be able to create a feeling of ownership and involvement by all; the FI Group which was mentioned earlier is one that seemed to be able to achieve this over its first few years, and even now, after flotation, around half of the equity is still held by staff. All these qualities within a company have to be acceptable to the market, but in fact many of them are just the things that the market is looking for in a successful consultancy organisation. They give an immediate impression of the entrepreneurial and dynamic nature of the company, as well as one of reliability and consistency; significant staff ownership in a company is also an added attraction for clients - *'their contact has a stake in the company and therefore won't allow others to let the client down'*.

In addition, there is the need for constant communication and the involvement of everyone; the old adage that *'the company that talks together, stays together'* is very true for this situation. Loyalties need to be on Handy's 'twin- track'[9] of one to the local operating unit and one to the centre; they should not be diverted elsewhere, and certainly not to any intermediate layer, which is often artificially imposed to *'improve management'* and cohesion, but which, in effect, usually adds cost and confusion.

6.4.2. Attitudes of staff

In considering the culture within which staff work, it's also necessary to look at the inherent attitudes that staff have to costs, cost reductions, performance and efficiency improvements. In the past, these have always been sensitive and threatening issues.

In Chapter 4, I talked of a culture in which responsibility for personal development now rested very much with the individual concerned; where this is generally accepted, by both staff and the company, then staff will almost

---

[9] The Empty Raincoat, Charles Handy. Arrow, 1995

automatically be looking all the time at cost and efficiency improvements. Then, as they keep identifying these, there will always be improvements in performance and, usually, a follow-on job for themselves. These kinds of changes represent a move from outdated 'command and control' attitudes of management and staff, to one where individuals feel personally accountable for all that they do, and may well report to a number of people simultaneously.

There therefore needs to be an attitude by everyone in the company of:-

- is this operation necessary?
- can we do it another - i.e. better - way?

This process has to be ongoing and cyclical to meet the ever-increasing demands for better, yet cheaper, services in the global market-place. And it needs to be initiated by the staff themselves rather than imposed through outside experts, e.g. other consultants.

6.4.3. Participation and rewards

In the new, or renewed, model of engineering consultancy, where innovation, co-operation and constant development are watchwords amongst all staff, what is equally important is that staff, somehow, also participate in a meaningful way in the success, or difficulties, of the company, other than in the most extreme way of simply having a job or being made redundant.

This is a concept that engineers may not be altogether used to!

Companies need to make a profit to live - everyone knows that. They also need to make profits so that they can go on doing things better and thus develop and survive. Profit is not an end in itself; the profit has to be used for the good of the company.

And the staff must somehow also share in that profit. So the way in which people are paid becomes very important. Traditionally, remuneration within firms of Consulting Engineers, and indeed in most other industries, has been by way of a fixed salary, whatever the actual performance of the company or business unit, with perhaps a modest bonus if the company has done well; the bonus is thus a share of the profit that has been made.

However, these bonuses have generally been small in relative terms and they have not therefore really been an effective incentive to generate profit. A different way of structuring salaries, following Handy, might have 4 components:-

- basic salary
- profit share in the total company performance

- share in the value added by the individual's own work unit, local office, team
- personal bonus for individual contribution.

With this type of arrangement, loyalty is preserved; however, the bonus elements need to be large, perhaps getting on for as large, in total, as the basic salary element. Only when these kinds of proportions are achieved will the system be effective. Two of the criteria which need to be applied to every single job in a company in this scenario are:-

- what do we pay for?
- what value is this job supposed to add?

But this will need a sea-change in attitude, since it will require a complete regrading and re-appraisal of everyone in the company, because the basic salary in the future is likely to be considerably reduced, perhaps even half, from what is the level at present. This principle has been applied for many years in a number of Japanese companies, and it has undoubtedly contributed to their prolonged success and their stable and committed work-forces; the recent downturn in economic activity in Japan has its roots in the financing of industry rather than in the way in which the work-force is employed.

An alternative is to try to adopt the 'team reward' approach - a difficult, but possibly very appropriate, way of committing people who work in team structures, as most people within consulting engineers find themselves. *"Team reward works well as long as you target well"*[10], so the skill will be to set and be open about easily measurable results.

All these kinds of changes are extremely difficult to implement; above all, they need the confidence and the trust of all, and that is where the work must start. But then, in a caring, communicating company, trust is not a problem!

6.4.4. Working practices and stress
Alongside the reward and cultural aspects of employment in organisations that are going to move forward, there is a need to find a balance in the way that the growing demands of work are invading the personal lives of all staff.

---

[10] Marika Szalanci, Compensation Manager for Rank Xerox, the Guardian, 27th April 1996

This is a concern for any company, as excessive stress amongst its staff and an imbalance between work/leisure/family/social life will result, long-term, in decreased commitment or effectiveness on the part of each employee.

Companies *"are now realising that they have to pay attention to - and respond to - the pressures which their people are facing in their working lives"*.[11] This is all part of a cultural shift and a change in the norms and values which have traditionally regulated behaviour at work.

Of course, in work within consultancies, where many, if not most, projects are undertaken on a team basis, the team effect can help to provide personal support to all of its members. Each member is, in fact, dependent on others in order to get their job done properly and hence make a satisfied client; there is, therefore, a mutual interest in ensuring that no single team member fails through 'burn-out'. Similarly, where several teams, perhaps even from different companies, are working together on a project, there is always the need for mutual respect and, if required, support, when things get tough, for one particular part of the overall team; otherwise the project suffers, and all those who work on it.

In the same way, each team becomes virtually self-accountable for its own   performance. Traditionalists may blanch at the dispersed and loose nature of project teams with their constant comings and goings as the project moves through its various stages, but, in practice, within a 'good' team, motivation amongst all the members ensures that accountability, meeting deadlines and work quality are all high on *everyone's* agenda. Nobody in the team likes to think of themselves letting the others down, but teams do fail from time to time, just like individuals, so a supportive environment is vital!

6.4.5. Attitudes to change
Change is now a way of life for business - things move on so quickly. That's why everyone in the company needs to identify problems and then pursue change themselves; they will be the ones to see the problem first and then they will change first.

With this kind of attitude, the people who are in continuous change will then clearly become the primary source of competitive advantage for the company. These people need to be held on to, and then the company can be developed and built on them, and the more of them, the better. If this kind of attitiude becomes the norm, then the risks to the company become lower and the mistakes fewer; the company gets used to 'running ahead of the flow'.

---

[11] The Quest for Balance, Roffey Park Management Institute, 1998/9

All companies need to build this culture, yet it must sound risky to management and, to some extent, it is; and it will take time. But when people feel involved in a meaningful way, then they do not only easily go along with change, it is they who initiate it.

6.5. SOME ELEMENTS OF A FUTURE STRUCTURE

These aspects of strategy, scope of the market and the skills and employment conditions of the work-force provide us with an initial guide to the way in which an engineering consultancy organisation can be structured. This will be based on a number of principles for the company:-

- it should not be hierarchical
- it should be set up as a network
- elements of the network should form the corporate whole
- loyalty should only be 'twin-track', i.e. no extra layers
- every individual should 'belong' to a recognisable unit
- delegation and team-work must be the motivating factors in all projects.

These may not be new concepts, but many of them have been lost sight of, in my view, over the past few years. These older, more 'human', arts of managing people need to be rediscovered, I suspect.

6.6. TAKEOVERS AND ACQUISITIONS

No discussion of strategy can be complete without serious reference to this topic, for, in almost all fields of business, there has been, throughout the '90's, an extensive amount of take-over/merger/ acquisition activity.

These moves have usually been justified by such things as the need to build larger market share, improve margins by eliminating duplication, or create an ability to offer a more comprehensive service to clients. Often, these are good and sound reasons for such a strategic move.

However, successful take-over/acquisition activity, in my view, requires a lot of careful planning and research; it needs much more than the two Boards of Directors agreeing with each other that a take-over is a good idea; it should also require much more detailed investigation than the conventional accountant's 'due diligence' reports. For, above all, any such move must look into the future for the new, changed, company; this is always difficult, can rarely be quantified, and cannot usually be hived off to other professionals, and given credibility by their 'Indemnity Cover', however diligent their researches into the past might be!

Success of a strategy of acquisition requires some basic ingredients:-

- the culture of the acquired company must be in sympathy with that of the acquirer. Rather than force two companies together that have different cultures, it is better to start a totally new company that has the acquirer's culture, but build it by using blocks from both. Never hesitate to ditch blocks which get in the way; that process is often more effective in the long run than trying to change them.
- the work profile of the acquired company should be complementary to the new parent, either in discipline, clients, or geography.
- the profile of the acquired company in the market must be sufficiently distinctive that the new parent can build upon it.
- the acquisition must provide opportunities for cross-fertilisation. The acquired company should not resent the take-over.
- continuity of personnel, reputation and client contacts must be presumed; this may require careful identifying for the new group. If all senior management from one company are likely to depart, then there's a need to carefully assess the value of any acquisition; it rarely proves possible to 'buy' work-load in the long term.
- most acquisitions are take-overs, yet managers often try to dress them up as 'mergers'. Most of the staff involved see through this, and it discredits the whole process and destroys trust and confidence. If it's a take-over, say it's a take-over!
- Accept that the acquiring company may change just as much as the acquired company; after all, the acquired company has been bought because the acquirer appreciates something about it.

And all acquisition or merger activity will need a lot of subsequent working at, at all levels; growth in this way often leaves a lot of loose ends trailing in its wake, a recipe for unhappy clients and subsequent business problems, not least of which is staff morale. Roffey Park research shows that the initial restructuring and changes do not, of themselves, produce new and beneficial working relationships; these can only be established by managers at local level co-operating with one another to achieve the purpose of the acquisition and to create integrated teams for the new organisation.[12]

---

[12] Developing people, Roffey Park Newsletter, Autumn 1998

A further suggestion for the immediate aftermath of a take-over - *"do not let anyone from the acquiring company visit the newly acquired company for at least two months; encourage any number of visits the other way round. Enforce this rigidly. This starts the process of building trust and confidence"*.[13]

NOTES ON THE CHAPTER

---

[13] Seminar address to Foundation Teambuilding by John Harvey-Jones, October 1995

CHAPTER 7

The organisation of engineering consultancies in the future

I have looked at how engineering consultancies have tended to develop in recent years and at how the markets of the future need to be satisfied. I have also looked at the different ways in which companies are beginning to operate in response to changed expectations amongst their staff, and (often) stakeholders, and to the changed employment conditions which now prevail and which will, undoubtedly, change still further in the future. All of these elements lead to the need for:-

- new strategies for engineering consultancies; these were examined in Chapter 6, and
- new ways in which Consulting Engineers should organise themselves so as to address these strategies effectively, and thus achieve the required performance and profitability needed for survival and viability.

In this chapter, I look at how Consulting Engineers should organise themselves to meet all of these requirements, as it's my belief that this is often at the root of the industry's difficulties; I feel that inappropriate organisational structures and management aspirations can often severely hamper the development and growth of viable companies in this sector of industry.

7.1. MANAGEMENT AND ORGANISATIONAL STRUCTURES

7.1.1. Types of structures

The overall objective of any company must be long-term survival - longevity. It would appear that it is those companies, in general, who concentrate on asset return, rather than longevity and continuity, who are shaken up, taken over, or fail most frequently; this perhaps can be reflected in the difference in prevailing attitudes within Anglo-Saxon/American organisations with their concentration on achieving the right financial results in the relatively short term, as opposed to the more European, or even Far Eastern, concentration on continuity and longevity. Europe's longest standing company is a Swedish company, Stora, formed originally more than 700 years ago and still trading, albeit in a derived field from its original activity. Shell is another good example of a stable and long-standing company, founded in the last century, whilst, of the top 100 companies in the FTSE index in 1965, only half were still free-standing companies some 30 years later.[1]

In the developed world, it would seem that the average life-expectancy of a company might be 40-50 years, i.e. about the same as the working life of its people! The characteristics of greater life than this or longevity are probably, and in general,:-

- conservative financing - i.e. often relying primarily on their own money
- sensitive to the world around them - i.e. broad-minded people
- consistent sense of cohesion and company identity
- survival, rather than profit, is the main driving force
- the company has changed successfully by using decentralised and delegated structures - i.e. there has been an element of power to the workers.

Interestingly, these factors do apply to a number of UK firms of engineering consultants; Halcrow, Mott, and Babtie spring to mind as representatives of the longer-lived and larger consultants who are still free-standing companies or partnerships.

Other characteristics of long-lived companies, generally, are that their leaders were **tolerant**, they let things happen and encouraged a management culture that always remained entrepreneurial rather than simply functional, a

---

[1] Unlocking the Secrets of the Long-lived, Caroline Glynn, Roffey Park Management Institute, 1996

culture where managers had '*flair and creativity*' as well as being well-organised. Whether the same can be said of the founding fathers of these firms of Consulting Engineers is rather open to doubt, so far as I understand! - but yet the firms that they founded have survived.

Within engineering consultancy, as indeed in most consultancy companies, client continuity and the stability of the client relationship are essential to at least maintain some continuity of workload. Thus their management and organisational structures need to be of high tolerance and flexible in order to create the right climate of viability and longevity, whilst also maintaining those other key elements within engineering consultancy - creativity and innovation. It is clear from the pattern of development of engineering consultancies, which I described in Chapter 3, that there has been much instability amongst leading firms in the profession over the last ten years; much of this can be attributed, I believe, to inappropriate structures, as well as the common problems of competition and economic conditions.

Looking, for example, at work within Europe, there are a number of engineering consultancies, and other service professions, which have developed comprehensive networks of offices across the EU during the last decade. These have generally had mixed success, for a variety of factors, but organisational structure is likely to be one of the important elements. The more successful ventures have relied on local operations being pre-eminent within their own markets, without the rest of the network being heavily involved; the benefit of the rest of the network has come in specialist task or project areas, as an 'add-on' to the local service. These ventures have therefore been based on the model of separate companies but all within a single federation. The less successful ventures have tended to work on the basis of a dominant main company, usually the original parent, with other EU offices being regarded more as simply branch offices; this hierarchical network then suffers from the fact that it tends to stifle innovation and creativity away from the centre.

Similar difficulties have been apparent with other service professions. Writing about the development of accountancy firms on an international scene, David Pitt-Watson, Director of Braxton Associates says, "...*Clients are beginning to require international service delivery. This cuts across the traditional structure and culture of partnerships....it requires team working across national boundaries, and therefore planned integration of resources. It is not always easy to get (individual national) partners to go along with this.*"[2]

---

[2] Exploring Corporate Strategy.Johnson and Scholes.Prentice Hall, 1993

A key observation regarding the provision of engineering services is that it is very much the people at the point of delivery, i.e. those in contact with the client, who will determine the competitive advantage and commercial success of the company; these people are often remote from the control of central management. Thus the structure of the company has to be loose so as to allow the client-contact people to further the client relationship in their own way, but to the maximum benefit of the company. Within engineering consultancy, almost all work is project-based and usually one-off; there needs to be a lot of up-front marketing to secure work in the first place and then the need to stay close to the client for anything else as follow-up work that he/she might have in mind, albeit perhaps in a totally different field.

The essence, therefore, of any successful service organisation is how its people, its workforce and its service deliverers are organised. This is clearly more complex than it appeared at the outset.

For a while, matrix-type structures appeared to be a solution to many of the difficulties that large international firms were experiencing; the same ideas were also taken on board by domestic consultancies which had extensive networks of offices across the UK. Matrix organisations, by their intrinsic nature, often have 'blurred' management and responsibility chains, whereby managers report both centrally and locally; they rely, for their effectiveness, almost totally on good internal relationships and a genuine spirit of co-operation and a willingness to share between employees in different locations and with different technical skills and/or client bases. Many of these kinds of structure in the past did not survive for long; the pressure to achieve short-term savings and better financial performance destroyed the principles of internal co-operation across the different companies, and the fact that these were often of a different size and profile to the original parent did not help. Internal competition is always the enemy of matrix-type structures, which rely on good, open, honest and trustworthy communication and co-operation between the various parts of the organisation; indeed, it is to be wondered, at times, whether such organisations can ever really work effectively within our British culture with its inherent spirit of suspicion, confrontation and competition.

Possibly the problem was also that these forms of structure were introduced without any training/preparation or consultation with the people concerned about how to work effectively within them.

But Consulting Engineers are not alone with these difficulties. KPMG found that client expectations began to change very significantly in the 1990s. Clients began to expect their advisers to fully understand all of their problems and to then provide a complete range of services to resolve them.

The firm had previously been very discipline focused; now it needed to become much more client focused. The skills, behaviour and attitude of staff needed to change to reflect this new emphasis and the control and measurement systems had to be adapted to the new culture. A lot of work was done to try to achieve this, but the firm still found that its people found it very difficult to face two ways, being on the one hand discipline and skill orientated, whilst on the other being able to respond to the client's full range of requirements. At the same time , the internal control systems still reflected the wrong emphasis; everyone found the new way frustrating and negative: the new matrix did not work. Consulting Engineers have also found very much the same difficulties.

An alternative approach was tried by ABB. *"We are not a global business. We are a collection of local businesses with intense global co-ordination. This makes us unique,"* said Sune Karlsson, head of the ABB Transformers Business Area in Mannheim, who could draw on the resources of 25 factories in 16 countries for his business. Within ABB, it is the Business Area Heads who coach, cajole and encourage the profit centres to co-operate with and lean on each other; the Business Area teams are always small, often only a few people, but they have to be people with a very wide grasp of what they are seeking to achieve for their customers. These people have to have a deep understanding of the inherent conflict points between the different parts of ABB and of the people within those parts, so that both business development and human relationship skills are vital personal qualities for them.

Thus, there are 3 key factors in organising a large and/or international business such as engineering consultancy:-

- an organisational structure which produces internal synergy
- a structure that encourages co-operative internal alliances, as well as external alliances with suppliers and customers, or even with competitors in the interest of keeping close to a particular client
- a structure that maintains a steady flow of new ideas about new products and new ventures for the company as a whole.

In their book, "Managing across Borders,"[3] Bartlett and Ghoshal examined the relationship between the nature of a company's trading environment and its organisational structure. Whilst their comments were

---

[3] Managing across Borders - The Transnational Solution, Christopher A. Bartlett & Sumantra Ghoshal. Boston: Harvard Business School Press, 1989

114

aimed primarily at large international or global businesses, many of their thoughts are appropriate also for large national consultancies, with their many different office locations and varying skills, as well as for international engineering consultancies.

Bartlett and Ghoshal identified 4 broad types of organisation:-

a. International Divisions with a strong home-based culture, highly centralised, and with technology transfer primarily from the centre to the foreign, or distant, unit. These structures work well where there is a range of foreign businesses, but where the business skills and types of work are relatively narrow. They never respond well to differences in local market requirements.

b. International Subsidiaries, or National but Distant Subsidiaries, who are very much responsible for their own destiny. This allows local management to take virtually full responsibility for their element of the business; the control of the parent is likely to be limited and to be dependent on co-operative reporting systems. Global co-operation and co-ordination are likely to be low, other than on a 1-1 basis with the parent or main company, so that the benefits of the single overall organisation operating worldwide but offering different skills from different locations tend to be reduced.

c. Separate Product Companies, whether globally or nationally; ABB is organised a little like this. There would be difficulty in managing multi-discipline work for an engineering consultancy with this type of organisation, and it would probably have exorbitant overhead costs. This is the model that large industrial Japanese companies have used with success, but there is a tendency for local touch and responsiveness to be lost.

7.1.2. The Transnational Consultancy.
This is Bartlett and Ghoshal's 4th type of large company organisation which attempts to combine the local responsiveness of the international subsidiary with the advantages of co-ordination found in separate product companies. The key to success with these types of organisation lies in creating an integrated network of interdependent resources and capabilities. Johnson and Scholes[4] comment that organisations like this would have these features:-

- Each national unit operates independently, but is a source of ideas and capabilities for the whole organisation.

[4] Exploring Corporate Strategy, op. cit.

- National units achieve global scale through specialisation on behalf of the whole organisation.
- The centre manages a global network firstly by establishing the role of each subsidiary, and then by sustaining the culture and systems to make the network operate effectively.

Clearly, for any large multi-disciplinary engineering consultancy, this is the type of organisational framework which is likely to lead to the most successful long-term results. It is a structure that, if managed and maintained properly, can survive change in the industry and which can grow new skills as they are required in the most appropriate location. It does require an element of 'balance' between the different national or production units and, like all matrix structures, it depends on the willingness of everyone in the organisation, and this means a real willingness not just a notional one, to participate and contribute to the well-being of the whole.

The idea of each separate unit having a speciality, or 'centre of excellence' role, for the whole of the organisation will do much to maintain overall motivation and the need for all parts to talk to, and work with, one another. It is when that particular speciality declines in importance, or becomes more universally undertaken, that defensive mechanisms set in on behalf of the people in that particular speciality; then co-operation and openness break down.

Figure 7.1 summarises the differences between the various types of organisational structure.

		GLOBAL	**CO-ORDINATION**
		LOW	HIGH
LOCAL INDEPENDENCE/ RESPONSIVENESS	LOW	International divisions	Global product companies
	HIGH	International subsidiaries	Transnational companies

Figure 7.1 *Organisational Structures within large Consultancies*

Clearly, the form that a large engineering consultancy takes in the future is dependent on many complex factors, not least its perception of:-

- the market, now and in the future
- the technical skills of its people
- how the company wishes to grow and/or change, and,
- most importantly, the personalities of its management.

Primarily, it is management personalities, and also the culture and personality of all employees to some extent, that will determine the success of the form of organisational structure that is finally adopted; this particularly applies if the Transnational form of matrix is adopted.

Whatever final form of organisational structure is eventually adopted - only, no doubt, to be changed again before long! - the key would seem to be to '*empower*' local managers and allow them to be entrepreneurial and pro-active with clients within some form of decentralised structure.

Designing and establishing this new form of organisation is far from easy. As Handy says,

"We used to think that we knew how to run organisations. Now we know better. More than ever they need to be global and local at the same time, to be small in some ways but big in others, to be centralised some of the time and decentralised most of it. They expect their workers to be both more autonomous and more of a team, their managers to be both more delegating and more controlling."[5]

Above all, with this type of structure there is the need to avoid inserting additional layers of management to improve co-ordination and co-operation between units. This is not only costly, but it usually ends up by creating a frustrating task for the extra layer and generally confusing everyone about their own role. It is far better to develop the company's employment and personnel policies so that everyone in the organisation feels that they want to co-operate with one another and will 'go the extra mile' to help a colleague or sister unit out.

All of this requires extensive effort in training people for working within the organisational structure and frequent people interchange so that key staff become familiar with one another. All must also remain flexible and adaptable to change.

---

[5] The Empty Raincoat, op. cit.

7.2. EMPLOYMENT POLICIES

7.2.1. The people dimension

In most countries, nowadays, engineering consultancy is a very competitive industry and there are few impediments to staff mobility. The correct choice of organisational structure, by itself, will not ensure that all goes well with the company; the attitudes of everyone who works for the consultancy will be important and, in this, employment policies, motivation, reward systems, perhaps even share ownership and stakeholder involvement, all have a part to play.

The 'people dimension' of a company has been likened to its intellectual capital, with the comment that in nearly every organisation today, *"the value of its intellectual capital represents more than 50% and sometimes over 90% of the total"*[6]. In enginering consultancy, a figure nearer the upper end would seem to be appropriate; employment arrangements are therefore of paramount importance for a company's future.

In Chapter 4 I looked at the changing world of work and employment as it applied to almost every industry, and in Chapter 6 I set out some of the issues which need to be addressed in an effective employment policy.

Within many industries, there has been a shift away from hierarchical structures towards a more open culture of relationships between people at work, one that encourages co-operation and a general mutual helpfulness, alongside individual responsibility. In developing that sense of responsibility for their own work, many people have also taken on a greater responsibility for their own personal development. Certainly, this is much more the case now within the UK consultancy industry, for whereas in the 1970s and even 1980s, employees could rely on their employer to provide a reasonably predictable career structure, this is no longer the case and it is very much the responsibility of each employee now to ensure their own personal development. Moss Kanter has called this *"the development of a post-entrepreneurial career mentality"*.[7] She illustrates how anxiety grows about the future, how this is translated into a constant concern about the next job for the employee, which automatically demands excellence in the present one, and how the allegiance of the employee can thus switch away from the present employer to the client for whom he or she is working at the time. In many situations, this switch can also be helpful to the current employer,

---

[6] Statement by Andrew Mayo, Professor of the Middlesex Business School, 1999

[7] When Giants learn to Dance, op. cit.

because the company also wants to ensure that their client base is content with the service provided.

For many people, this concept is quite a challenge. But they need to address it, just as much as engineering consultancies need to encourage their people to think this way. They need their people to be successful, full of innovation and creativity. *"People who are uncomfortable in an unstructured world, don't make it,"* was a phrase once used by Dick Liebhaber, Head of Operations at MCI.

So alongside organisational change within any company, there is a need to help the people who work for it to feel comfortable, as well as challenged, in their work. One of the ways in which this can be achieved is by the employment policies which it adopts and through the opportunities for retraining and changing which it provides.

The construction industry, as a whole and in the UK certainly, is notoriously bad at providing training for its people, partly through the short-term pressures on results, but also, I suspect, through sheer lack of foresight and imagination about the long-term consequences of having a work-force that is not prepared for ongoing change, both technically and managerially.

In the longer haul, there are several key features to instil into people who are working in a transnational form of consultancy:-

- a spirit and an ability to co-operate internally within the company and externally with clients and peer companies on a project,
- an ability to co-operate even with traditional competitors within a project relationship,
- a feeling of being involved in, and a part of, the future of the *whole* company.

This latter feeling can be induced in a variety of ways, perhaps through shareholding or being a stakeholder in the business in some way, through a Workers' Council, thus giving access to top management on key decisions, or, more simply, through a very effective and personal internal communications strategy, with allowance for feed-back and influence. Whichever approach is adopted, the key idea to convey is that management is taking medium/long-term views of the future; this helps to reassure people that they have a future within the company, that their contribution is valued and that, in some way, they have an element of ownership.

Somehow, the objective must be achieved of projecting an image of scale and size in the engineering consultancy market whilst retaining that personal, individual and perhaps small link to the client and to colleagues. The concept of twin citizenship seems to me to be a useful one in looking at employment and personnel strategies that fit this objective. Everyone in the company needs to have a local loyalty to their immediate colleagues; deny this and one kills all incentive and instils reliance on others somewhere else. At the same time, there has to be a loyalty to the bigger organisation, for without that, misunderstanding, jealousies, duplications and inefficiencies will all flourish. With these twin loyalties must also come an individual approach to flexibility and an ability to cope with a myriad of simultaneous tasks, for such is the character of today's world. These all mean that there is a great need for personal development and for people to be happy and content with just '*who they really are*' rather than concentrating on what they might aspire to or what title/role they might hold. This means that people need to be caring as well as tough within themselves, just as much as they are with others. In an industry that, certainly within the UK, is dominated by men, it's ironic that these are very much traditional female qualities!

7.2.2. Internal structures
In Chapter 4 I portrayed the internal structure of a company as having 3 levels - the 3 leaves of the 'Shamrock'. Whilst many of the above aspects are vital for the 'central core' of staff, many of these should also apply to other categories of staff, certainly to the 'contracted fringe' who are permanent employees in all but actual payroll count. This all requires a high degree of management sophistication; no longer will engineering consultancy be able to operate by means of a simplistic 'command and control' structure.

Management will need to be sensitive as well as persuasive, encouraging as well as directive, liberating as well as controlling.

7.3. TEAMWORK
As work processes for Consulting Engineers become more complex and multi-disciplinary, so the need for team-working will grow. These teams will need to be effective teams rather than simply groups of individuals with different skills who work on the same project or task at the same time.

Some of the intrinsic characteristics of effective teams are that the members listen to one another, work inter-dependently and are prepared for constant change and evolution as the expectations of the team and its skill composition change.

Teams need regular reviews to see how they are doing; these foster teamwork and build up trust, momentum and confidence, Such reviews will also throw up changes in targets and work requirements; if extra effort is needed to achieve these, then a sound team approach will help everyone to 'go the extra mile'.

And team reward systems, as mentioned in Chapter 4, will help with all of this. A significant element of team reward, set alongside existing individual pay systems, will be needed if the team is to be effective. This kind of approach will help to take out individual competition within the team and reinforce that feeling of mutual interdependence.

Such interdependence is tested at times of stress, or even failure, for the team; a company's attitude to failure must be one that enables it to be seen as both a learning experience for all involved and a strengthening of bonds within the team.

Like many aspects of management and organisation that I have already illustrated, teamwork has to build on the creativity of each individual member within a team so as to obtain an optimum result; yet, at the same time, each person needs to feel confident in their own right about their place and their abilities, so these all need building up as well!

7.4. TRAINING AND DEVELOPMENT

Alongside all of this, there is the constant need for maintaining and updating technical skills; this is often a personal task although any employer who neglects this aspect is likely to find himself left with a workforce that has no longer any 'state of the art' skills. Particularly at managerial level, there is a great need for training to improve management performance. As Sir Alan Cockshaw, Past-President of the Institution of Civil Engineers, recently wrote:

"There is nothing more important for the future of the construction industry in this country than the quality and professionalism of its people. The industry has to be looking for a programme to help senior people with business management - as distinct from technical management."[8]

There are some signs that engineering consultancies are beginning to face up to the need to develop their staff in this area. In a recent staff development programme designed by outside consultants for the senior management of DHV, one of the largest Dutch consultancies, the items covered included:-

- Strategic issues, visionary thinking and the 'world around us'
- Commercial skills and business acumen

---

[8] Director Development Programme, by Thomas Telford Training, 1999

- Change potential and willingness to change
- Motivation and leadership skills
- Conflict management techniques
- Personal development and risk taking
- Client and partner relationships
- Internal needs and co-operation
- Communication skills and information strategies.

The programme went on to devise ways of assessing management potential and of ensuring that senior people were aware of their limitations/areas of excellence, thus enabling the company to be organised in terms of its senior staff so as to maximise their skills and potential. The exercise, as a whole, was intended to ensure that the consultancy felt comfortable with exactly where people sat within the organisation and that there would be an increased willingness on the part of everyone to co-operate with one another. In common, perhaps, with many exercises like this, it was a difficult programme to develop and run at the beginning, but it has apparently now proved to be successful, with many senior staff having participated.

This is the kind of approach that is also needed for the 'Latham' ideals of partnering and co-operation within project teams to take off; DHV's programme was an attempt to change cultures within a particular company, but the same kinds of change are needed across a whole industry if the benefits of partnering, etc. are to be properly realised. As one commentator has written:-

"Cultures and personalities can be changed but the process is difficult and slow, requiring an understanding of the social and psychological pressures at work. The values, beliefs and principles held by members of an industry are inherently hard to change because of being absolutely true and 'obvious' to those who hold them. People tend to resist change so that new industry initiatives are viewed with caution, scepticism and political bias. In order to mould an existing culture, initiatives are required which go beyond the superficial level of behavioural changes and strike at the deep-seated values and beliefs which subjectively control the responses of members....The remedy must lie in the training of the next generation of young professionals. A set of core values must be taught.. .....appropriate attitudes and modes of co-operative professional behaviour."[9]

---

[9] Len Bird, writing in Civil Engineering, Institution of Civil Engineers, February 1997

Overall, however, commitment to R & D within the UK construction sector as whole is low, by international standards,[10] with total R & D spend being around 0.5% of the turnover of construction. Less than a quarter of this, so around 0.1% of the total, is invested by companies in the sector themselves, and these are primarily the materials producing companies in any case rather than contractors and consultants. The figures probably exclude quite a lot of specific staff training and development, but it still seems likely that the total spend on training, staff development and research is seriously insufficient for the changes in technology and working practices that need to be achieved.

There are signs that many companies are investing quite heavily in new I.T. and technical systems, but there would appear to be much less investment in people training, particularly on the management side. Training in management skills is essential if engineers are to continue to remain the main players in their own business, rather than the 'ubiquitous' accountants. Many accountants have worked or been trained in a variety of companies, with an emphasis on running businesses; this has made them used to taking the lead and to looking at new ideas for businesses. In contrast, engineers have tended to be more channelled in their work settings; this makes them *"great middle managers, but poor top executives"*.[11]

A further thought is this. Once companies of Consulting Engineers begin to recognise that their only real assets are the skills of their people, rather than just the, narrowly-based, accounting definitions of fixed and current financial assets, then they will recognise the need to invest in them just as much as they need to invest in their other assets, like premises, computer systems, etc. This investment, which should be at the heart of their business, is rarely given a thought in public utterances; for example, the 'Placing Document' for one of the UK's largest firms of Consulting Engineers, W.S.Atkins, designed to impress potential investors with the value of the company, contained not a single word about the asset value or intellectual capital of its staff.[12]

The only real value that any consulting firm has is the intelligence, the integrity and the quality of its people. It is the asset value of the right kind of

---

[10] Research Focus, Institution of Civil Engineers, August 1996

[11] Engineers in Top Management, Institute for Employment Research, Univ. of Warwick, 1997

[12] W.S.Atkins plc, Placing sponsored by Schroders, July 1996

staff and expertise that will give a company a much more effective source of business advantage than any amount of impressive financial data and strong balance sheets.

Improving the asset value of a Consulting Engineering company thus requires real and continuous investment in the training, motivation and development of its people.

NOTES ON THE CHAPTER

CHAPTER 8

Business processes

8.1. BUSINESS PROCESS ENGINEERING & RE-ENGINEERING

Much has been written on this subject in recent years, showing how a systems approach applied to a range of business processes can achieve client satisfaction, i.e. the goods or services supplied comply in every respect, cost, delivery and quality, to the customer's requirements. Primarily, the technique has been used within the industrial and manufacturing sector where the element of repetition ensures that a change in approach is applied consistently.

Within the service sector, where usually each commission has different client and consultant parameters, it is more difficult to achieve such a systematic improvement; it should be possible, however, using the introduction of at least some benchmarking, comparison of internal work processes and (perhaps) comparison with competitors, to adopt the principles of BPE.

Comparison with competitors is, at best, difficult owing to the wide variety of projects and tasks within construction, but it may be possible to achieve comparisons for certain aspects of work. What will be much more appropriate for consultants will be to establish internal processes whereby, over time, the performance on similar tasks is measured, evaluated, improved and modified. This will involve the whole of the workforce, not just the project managers or accounting staff.

Benchmarking can help companies to really assess where they are; then the process of change can start. The whole concept has to start at the top and reach all levels of the organisation; there has to be a commitment throughout the company to establishing appropriate measures.

One example of benchmarking used by Carl Bro is the extent of I.T. used in project work; this is assessed year on year to track their ability to use available sources of information rather than for ever working things out afresh or 're-inventing the wheel'. This measure is also used to build up their cost base; for example, in 1998, I.T. costs were DKK 18,000/head, around 3% of revenue/head.

With the shift to partnering for projects and more work being done on a regular basis in alliances, these kinds of measures for assessing performance and identifying improvements will inevitably spread across company boundaries; they will also become essential elements in the steady drive to achieve the year on year cost reductions that Egan envisages for the whole of the construction process, including the consultancy element.

Drucker has shown that in every industry sector within the global economy there has to be continual ongoing improvement in costs and efficiency. BPE can be made to be part of this ongoing process, provided that its more measured, consistent and systematic approach is tempered with the encouragement of individual creativity and innovation. It is a tall and difficult order for management as Towill[1] has pointed out - *"All too frequently, top management has given little thought to the design of business processes;"* however managements which initiate this kind of thinking throughout their consultancy will be enabled to keep it competitive and thriving.

8.2. INTELLECTUAL CAPITAL

I mentioned, earlier in the book, the lack of emphasis on intellectual capital as regards engineering consultancy; for a whole range of service or professional firms, this element is an important one, yet so often the staff contribution to a company's performance is dismissed in the Annual Report with some platitude from the Chairman about *"their loyalty, motivation and high calibre."* And that's where it often seems to end!

Engineering consultancy is a business that not only has to be sound financially and commercially, it also has to be continually developing on behalf of its clients and be innovative so as to produce ever more creative solutions to clients' problems.

Thus, the concept of an *'Intellectual Capital Account'* has been created by Carl Bro, one of the large Danish consultants. This is an attempt to measure, year on year, the extent to which the company has *"managed to*

---

[1] Successful business systems engineering. Prof. Denis R. Towill. Engineering Management Journal. April 1997

provide an appropriate platform for future earnings".[2] Some of the measurements in this may seem somewhat mundane, e.g. the % of men vs. women and the age distribution within the consultancy, but all are then incorporated into an overall policy for the balance of energy, creativity, experience and sensitivity.

Carl Bro use 6 measures for their Intellectual Capital Account:-

- Human Capital - dealing with staff issues, training, satisfaction, etc.
- Customer Capital - measuring satisfaction, achievement of a balanced workload, as targeted
- Image Capital - assessing the 'image' of the group from outside.
- Innovation Capital - measuring new types of project.
- Process Capital - assessing the sharing of projects across disciplines and across the group.
- IT Capital - measuring the extent of the use of IT, etc.

All these are combined as shown in Fig. 8.1. Carl Bro are honest about their performance on these bases, e.g. *"the image result is not yet satisfactory".*

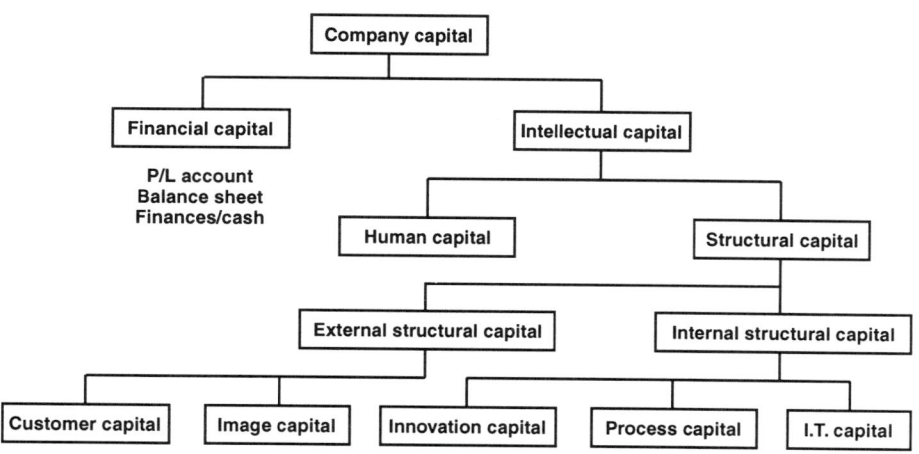

Fig. 8.1 *A Model for Intellectual Capital*

(Source - after Carl Bro A/s)

---

[2] Annual Report. Carl Bro A/s. 1997/8

This kind of approach is one way of trying to measure and record those elusive 'creativity, innovation and market perception' concepts that will be the foundation of long-standing consultancies in the future, rather than simply financial performance in the previous year. The latter aspect is, of course, of paramount importance to the shareholder stakeholders, but without the other concepts, performance will gradually decline and the staff, the mainstay of any consultancy and an equally important group of stakeholders, will 'walk'.

Whatever approach is used by other consultants, the key will be to be able to formulate these kinds of measurement using data that is already available within the management system, so that this aspect does not have significant additional cost.

8.3. CONTROL AND MANAGEMENT SYSTEMS

Every company will have its own control and management systems and these will probably be well established and will have been employed for many years. They may cover a variety of business factors, or they may be simple project control mechanisms linked into the overall financial control system for the consultancy. The key elements seem to me to be:-

- systems for testing the effectiveness of the marketing function, for developing business opportunities and for looking at alliance strategies
- systems for determining strategies for undertaking work, e.g. setting goals and performance targets
- project control mechanisms, in terms of resources, costs, cash and programme
- auditing of client satisfaction and project performance, so as to have some certainty about repeat business and perceived quality standards
- company control and forecasting, in terms of the Profit & Loss Account, Balance Sheet and level of financing.

All of these need to be linked into personnel and employment policies so that they are seen to be 'user friendly' and relevant to the business, for engineers are notoriously orientated towards technical excellence rather than business performance. It's also important to enable technical people to remain expert in their specific field and not to let them get bogged down with a steadily growing burden of 'busywork'; this can often be done by others who have different skills, *and costs*. 'Busywork' such as this usually makes little net contribution to a company's profitability.

The systems which look at performance and forecasts for obtaining work and for new business opportunities need to be set within the transnational culture of open-ness, without rivalry. There needs to be a constant awareness of the dangers of 'creeping market boundaries', for the *'grass is always greener on the other side'*; if boundaries creep, then this can cause different entities of the same consultancy to seek one another's customers; this leads to a culture of concentrating on defeating internal rivals rather than projecting the company as a whole against its external competition.

So far as project control is concerned, it is important that, at the outset of a job, it is quite clear whether the profit level that is projected is 'core' profit or 'development' profit, the former being associated with the view of the project as perhaps a 'cash cow', whereas a project of the latter type might be more of a 'star' project which will lead to new opportunities for the consultancy in the future, either locally or elsewhere. Projects should also be grouped by client, market sector, location, etc. so that the company can always be aware of performance trends within different markets and can incorporate these into future strategic thinking.

One of the well-known difficulties within engineering consultancy is the treatment at project level of 'Work in Progress'; this requires an assessment at every stage of the extent of the work that has still to be done, and the extent to which work which has already been completed, particularly if it is construed as an extra, will be paid for. This difficulty suggests that it might be worth trying to include within the project control and reporting system a wider range of performance indicators which will help to clarify or at least highlight these or other potential problems, such as:-

- client satisfaction
- continuity of project personnel
- relationship with partner firms and external authorities
- best/worse case scenarios for project out-turn in terms of resource input, delivery and quality.

The element of client satisfaction, or otherwise, is a particularly useful pointer to the state of the client relationship; this is a key ingredient for an assessment of the potential for ongoing business opportunities, as well as perhaps also being a pointer towards the possible outcome of any negotiation over enhanced project fees.

Monitoring of projects must also be seen in both positive and negative lights. In many systems, there is a tendency for only adverse indicators of project performance to be flagged up; this can be very demotivating to the staff involved, particularly if there are other pressures and priorities on the business which have brought this situation about.

Nowadays, much of this necessary control and monitoring information can be generated through appropriate IT systems without a lot of people input. This reduces and simplifies the management task and enables project staff to concentrate on the 5 main aspects of project performance, namely:-

- creativity in project conception,
- adherence to agreed targets and goals,
- maintaining client contact, satisfaction and quality standards,
- maintaining appropriate internal and external relationships, and
- working in such a way with the client that the next project follows on almost automatically.

This latter is an ideal situation, but one which Consulting Engineers must embrace if they are to avoid the continuous, costly, time-consuming and often frustrating task of forever seeking out new clients for the company's very survival, let alone expansion. Such clients nearly always have to be prised out of the clutches of a rival consultant, so it is never an easy task - so much better to continue to work with existing clients where there is some continuity of relationship.

8.4. THE INTERNET AND E-MAIL

The E-Systems revolution is changing not only the way that people react but, more, the way in which people think; thus the implications apply to all companies and all businesses. Suddenly things and concepts are within our grasp that, until now, have been regarded as 'way ahead in the future'.

E-commerce, Intranets and Extranets, now mean that so many more tasks can be done just as well at a distance; thus teams and individuals in several locations can work on a project simultaneously for a client that is similarly dispersed. The Internet is much more than a new technology for marketing and selling although, at first sight, this might seem to be its most immediate application; it is already a way of life for many companies. It is the start of a move to the *"virtual ideal in which customer, corporation, and supplier are indivisible".*[3] Suddenly, everything from new technology to market intelligence, both the very stuff of life to Consulting Engineers, is available at a touch of a button - and it's available to all!

But it's also important to ensure that, in spite of all of this electronic access, there is still a personal connection between client and supplier, and between partner suppliers. No amount of E-mail can be a substitute for mutual trust and confidence We have all had the experience of being answered by a remote answering service enquiring our business, instructing us to press star buttons, etc. and then the interminable wait whilst we get serenaded by 'Musak'! There is little substitute for a helpful, courteous, and intelligent response; without this, people businesses will surely suffer. It's very difficult to develop a meaningful client relationship over a computer!

This has tremendous implications for service industries where it is their relationship with their client/customer which is so often the key to successful business. It will encourage the trend, already identified in this book, for these relationships to be on very personal levels; it will probably appeal more to the younger element within engineering consultancy, thus enhancing the need for 'flat' structures, and will pose yet more alternatives for management to try to grapple with at the strategic level. It will simply exacerbate the need for more and more change if consultancies are to survive, for the newly-created competitors to the longer established firms will, without doubt, embrace the technology to the full.

This ability to communicate instantly, and across national boundaries, means that it suddenly becomes easy to establish small and very highly specialist teams in separate locations, or even within separate companies. This kind of development may start to lead to the break-up of larger consultancies, with their large offices and infrastructure, so that the core or heart of the organisation becomes simply a marketing, delivery and controlling function; then the specialist applications and technical expertise can reside elsewhere, possibly within supporting but separate companies which may work simultaneously for a number of competitors in the same field.

Thus, the present large company of Consulting Engineers would become primarily a project management company, whether at the project conceptual or implementation level, with a series of networked technical clusters doing much of the actual day-day work.

---

[3] 'Profit goes to the Swift and Lithe'. Robert Heller in Management Today. April 1999

8.5. FACILITIES AND PREMISES

With the focus also now having to be more on individual project outcome and with more flexible working patterns, there will be a shift in the office and facilities patterns for Consulting Engineers, as in many other service industries.

The multi-disciplinary nature of many projects means that it will no longer be appropriate for people to always simply work in a fixed place within their own discipline teams; they will work, possibly, within several different project teams at any one time, reflecting different skill mixes and types of project. For this way of working, offices and facilities will need to be flexible, open-plan type probably, with staff located in project groups for the larger projects, moving to new locations as the work changes.

These offices will emphasise internal communication and approachability, at all levels; there will also be quiet areas for private meetings and for 'thinking/breaking out processes'. A proportion of staff, probably somewhere in the range 20-35%, will only ever have desks on a 'hot desk' basis.

This way of working is common within many professions; the objective is to ensure that maximum value and creativity are extracted from teamworking, with minimum cost devoted to fixed and inflexible infrastructure, which itself can tend to inhibit co-operation.

The growth of the virtual office, using the present explosion in mobile communication and data systems, alongside the move to electronic filing and the 'paperless' office, is going to push office flexibility even further, with yet more savings in fixed costs.

And as the systems become more open, with most of the staff having access to them, so the culture of secrecy, 'the power of the information-holder', will diminish and a culture of collaboration will be encouraged.

Such collaboration will be essential in the dispersed office of the future. Already, within some consultancies, almost half of the staff have the option of some, if not all, of their work to be teleworked from home or from some other base; this proportion must surely rise.

8.6. THE MARKETING ELEMENT

Marketing for all companies is a significant element in company strategy and in developing the business; Consulting Engineers are no different. In fact, it can be argued that it is a much more critical function for firms of Consulting Engineers, in comparison with some other service industries, shown in an earlier chapter, the 80/20 rule of approximation probably still holds, so that the marketing focus must concentrate on the 20% of regular clients who offer the best prospects for workload sustainability and growth.

Too often, in the past, marketing by engineering consultancies has been seen as producing ever more sophisticated documentation about the services that a company can provide; this can be labelled as 'product-driven' marketing. What clients are really looking for is marketing which will help them to identify the best company to satisfy their own business requirements; thus marketing becomes totally 'client driven'. In client-driven marketing, the initial focus might be on 4 key elements that the client might want; these can be defined as client 'value' criteria:-

QUALITY   ~   SERVICE   ~   COST   ~   TIME

Table 8.2 sets out, in more detail, some of the elements which might make up each of these. Together, they will start to define the value which a client can put on what a company might be able to offer.

Quality	Service
- meeting customer requirements	- dialogue with client for project development
- fitness for purpose	- client feed-back and information
- minimum variation from agreed requirements	- ability to accept differing client requirements
- opportunity for improvement at every stage	- ongoing support after project completion
Cost	**Time**
- concept and detailed design costs	- delivery time for each phase of the project
- quality assurance	- ability to respond to market timings
- opportunities for shared savings in overall project	- lead times for mobilisation
- cost certainty for project	- coping with variations and effects on end-dates

Table 8.2 *Key Elements in Client Satisfaction*

Different clients will perceive some of these elements as having different levels of importance, but all are likely to consider these as critical elements; and so Consulting Engineers will have to address them specifically in their marketing approaches. For example, in the UK, it is likely that the private sector will value speed of delivery of a service, and the ability to respond to changes, as just as important as cost, since there is normally a need to bring a project to fruition as soon as possible, once a decision to go ahead has been made. In contrast, the public sector, with its need for accountability and disclosure, has usually been more concerned with cost as its first priority rather than speed of delivery.

Almost all of the items highlight the need for key staff; and indeed all staff in contact with the client, to have an ongoing and constant dialogue, both internally and externally. Practitioners involved in the project need to be seen to be doing this, rather than the more distant or removed 'marketing department'; this department is there more to provide the tools for developing the business, and for identifying future growth opportunities, rather than for actually securing commissions or 'closing sales'.

These tools will include:-

- brochures and project experience details -the 'glossies'
- similar projects undertaken, with project outcomes and, perhaps, client references
- staff experience and skills - not just potted c.v.'s, but perhaps team credentials and achievements, etc.
- training of front-line staff in marketing techniques, e.g. presentation and listening skills, understanding the client's business, identifying priorities, negotiating and closing sales
- providing personal support at critical points in the marketing sequence, perhaps when 'the going is getting tough'
- monitoring marketing activities, providing an overview and feed-back on these
- ensuring that individual initiatives and marketing exercises are in line with strategy. This will mean avoiding the temptation of pursuing leads that are outside the overall strategy, but which just seem a 'good idea' or a one-off opportunity which will not really take the overall business any further forward.

There is much to be said, in connection with this latter point, for having a system in place for assessing whether potential opportunities meet the strategic requirements of the business. In the successful business of the future, it will still be necessary to have a strategic focus on activities to ensure that the company does, and continues to do, *'what it is already good at'* rather than frequently or continuously embarking on new techniques and project technologies. Getting consistency into the work-search operation should lead to consistency at all stages of the business, and make the application of business processes effective.

In his book[4] a few years ago, Fisher set out the steps for a comprehensive marketing strategy and implementation plan for companies in the construction industry, including consultants. The approach outlined in his book, and the techniques to be used for assessing the potential of clients and opportunities, is methodical and worth studying. It is, however, a fairly elaborate approach for a fast-moving industry; it may not be directly applicable to the more spontaneous approach which is an essential element in business development within the consultancy field, where projects need to be identified at an extremely early, even embryonic, stage in most cases.

Another book, by Scanlon[5], might be more relevant for Consulting Engineers; this outlines a less formal structured approach to business development and illustrates some simple techniques for assessing project viability at the bidding stage, as is shown in Fig. 8.3. Although fairly simplistic in its application, this type of method is probably a more appropriate base model that marketing departments should use for their own particular consultancy in order to try to ensure that there is good correlation between the strategy for the company and the work-search operation.

Scanlon suggests that there should be a minimum score of 75 for any project proposal before it gets seriously investigated and pursued.

Above all, approaches to business development need, at all stages, to be in line with the strategic objectives of the company as set out earlier in Chapter 6. Thus, for example, involvement in a PFI project, perhaps with an element of performance or financial risk for the consultancy, needs to be squared with the availability of finance and the strength of the balance sheet. Many, perhaps more cautious but perhaps to be longer-lived, companies, take the view that, they will not participate in any risk that is outside their own control, thus limiting their role in these types of project.

---

[4] Marketing for the Construction Industry. Norman Fisher. Longman. 1986
[5] Marketing of Engineering Services. Brian Scanlon. Thomas Telford. 1988

All of these concepts are ways of overcoming the difficulty that almost all service industries face when they are trying to project themselves and achieve new business. Clients simply have difficulty, in many cases, in committing themselves to pay for a service; they are much more prepared to pay for results which they can see for themselves and which are of direct benefit to themselves. It is the task of Consulting Engineers, in their marketing, to demonstrate to clients where the service provided can lead and how the required result will be obtained.

Factor	Scores										Rating
	Positive			Neutral				Negative			
	10	9	8	7	6	5	4	3	2	1	
1. Relevance of past experience	• In-house			• Subcontracted				• New area			
2. Technical capability	• Proprietary leadership position			• Capable				• Needs development			
3. Team strength	• Best in field			• Best available				• Second team			
4. Support facilities	• Available			• Subcontracted				• Needs development			
5. Marketplace knowledge	• Active contacts			• Aware				• Did not expect invitation to bid			
6. Customer contacts	• Good working contacts			• Occasional				• None			
7. Competition	• Limited			• Open to wide number				• Obvious likely winner			
8. Customer expectations	• Can be exceeded			• Could be met				• Difficult to meet			
9. Commercial	• Can charge a premium			• Standard margin				• Needs cost cutting			
10. Relevance to future	• Supports a leadership position			• One of many such opportunities				• One-off situation			
										Totals	

Figure 8.3 *A technique for scoring bid attractiveness*

(Source - after Scanlon)

Overall, the work of business developers and marketeers, which many more people, in the consultancy of the future, will have to be skilled at, is to be a *'sparring partner'* for the client, as it was once put. The successful consultant will be the one who asks the really pertinent and difficult questions of the client in his or her attempt to really tease out the nature of the problem and the client's aspirations and capabilities. This process will build up confidence and trust between consultant and client; it all requires a high level of significant and vital commercial, negotiating and listening skills, together with lots of creativity.

NOTES FOR THE CHAPTER

CHAPTER 9

A holistic approach

9.1. A SYSTEMS APPROACH TO THE BUSINESS

The reader could be forgiven, at this point, for thinking that the creation of a satisfactory way forward for large national or international engineering consultancies is an almost impossible task; not only does it appear difficult and complex, but there is also no set model and no evident *'right'* way in terms of financial results and longevity. This is indeed the case, for there are many processes at work which provide, daily, fresh challenges to the established order of things within engineering consultancy.

Some of these will include:-

- developing countries are creating indigenous skills through improved education
- other professions are looking to add engineering skills to their own portfolios so as to be able to offer a more 'one-stop' service at a strategic/planning, or even implementation, level
- market pressures are changing the methods of contract so that consultants have to be able to carry additional financial risk - many may not be able to do this in the longer term
- the broadening of the scope of services that clients require is putting demands on staff to be even more innovative, responsive and wide-ranging in their approach
- staff within the company are always mobile -they may take secrets and client connections with them when they move on
- the Internet will introduce completely new ways of working and relating within and without companies
- staff within a consultancy are increasingly likely to want some stake in the company and in its direction.

140

These factors, and many others, are inter-linked, as shown in Fig. 9.1 below. They demand a systems approach to the whole of the business, one that is holistic in concept, recognising that the business is an entity that needs to be in harmony across all of its elements.

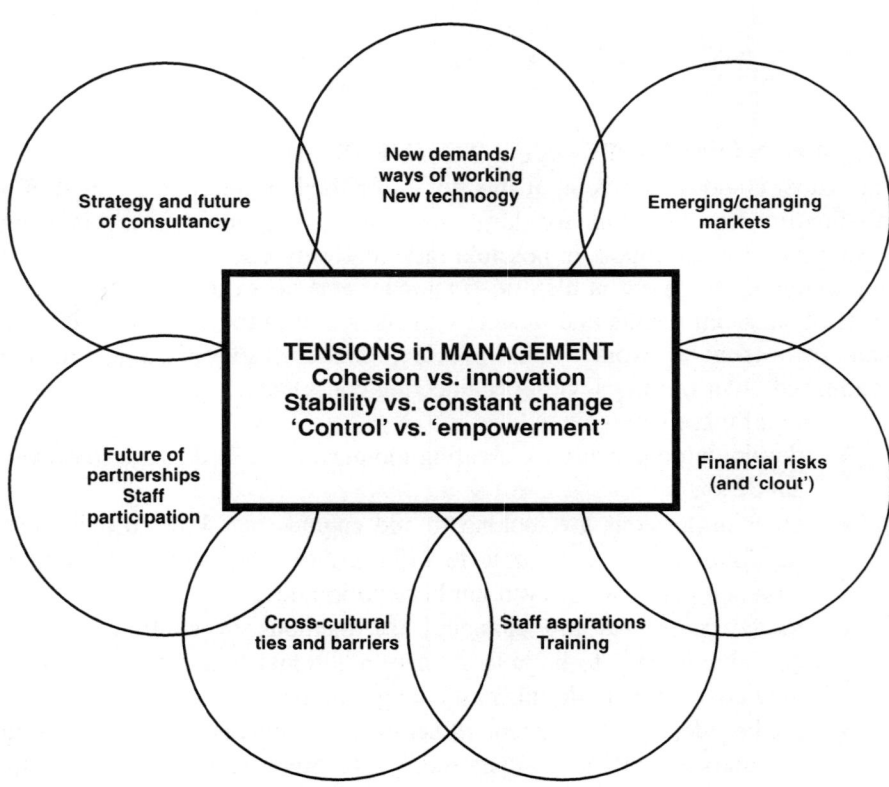

Fig. 9.1 *Overlapping Elements in Strategic Development*

9.2. HOLISTIC ASSESSMENT

All large Consulting Engineers will need to systematically improve their performance year on year, whilst continuing to strive for ever more credibility in a global market place. This will require balance between several different aspects, e.g. the employees, the clients/customers, the business, and the regulatory environment.

The assessment of intellectual capital that I described in Chapter 8 is one way of focusing on the employee aspects, and on client aspects to a lesser extent; but it ignores the business and regulatory environment within which every company has to operate.

Another assessment tool now being developed and used is shown in Fig. 9.2.[1] This has been developed by a management consultant to look at 3 key elements in a company's profile, first in isolation, but then assessing their overall 'balance' for the company as a whole. This type of holistic exercise highlights the likely outcome of a company's next move and, when compared to the present situation, enables management to assess whether the main thrust of change should be internally or externally focused.

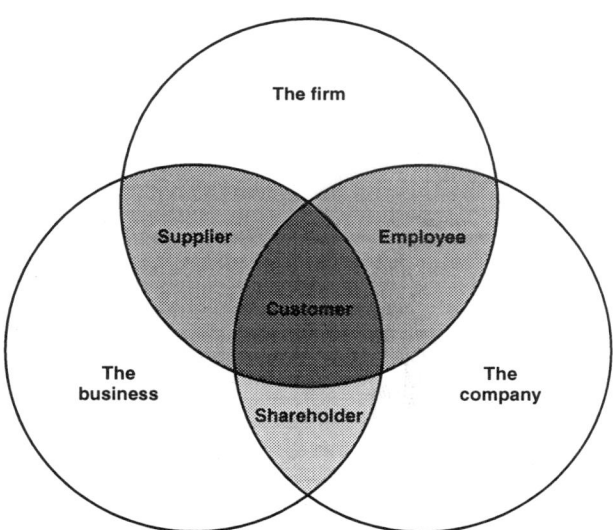

Fig. 9.2   *The Holistic Business Analysis*

(after Stratology Consultants)

---

[1] 'Springboard' - A Strategic Analysis Package. Stratology Consultants. 1996. See Notes to References

Essentially, these kinds of tool throw up imbalances within the business system; they thus provide a good starting point for corrective action. In this particular type of technique, there are 3 key elements (*sub-systems*) which should be in balance for a company to prosper:-

- The Business. The existing and potential market environment.
- The Firm. The Company's staff, their culture and way of working.
- The Company. Its systems, management and control processes.

These sub-systems have a dynamic interaction and are initially assessed independently of each other using a number of control parameters. The relationship between them is then tested and combined in the type of diagram illustrated in Fig. 9.2. This particular example indicates a company that is broadly in balance and well-focused on its customers, as shown by the large extent of overlap of the 3 circles.

But this type of tool seems to lack the defining relationship element between client and consultant, which is a characteristic of successful consultancy in any field, especially if the consultant is to be perceived as the friendly 'sparring partner' that I alluded to earlier. It is the quality of that relationship, right across all activities, which will enable the consultancy to be in the 'grow steadily and therefore survive' category, where the 'growing' element has to be not so much in terms of more and more disparate activities, but a greater and greater penetration into sophisticated technical markets, often inter-related to one another.

The holistic approach should therefore be at least a 4-way process covering:-

- human resources - qualities, quantity, aspirations, employment, etc.
- the market - trends, future scenarios, competition, technologies, etc.
- the clients - their aspirations, relationships, environment, etc.
- the business - systems, management, finances, etc.

Combining these factors and attempting to achieve a meaningful relationship between them is a complex task, but it seems to me that this type of analysis must be a useful exercise for any firm of Consulting Engineers. Even if the outcome of this does not prove to be particularly robust, the very act of undertaking something like this must be helpful in developing the people of the company, due to their involvement in this type of process. Then, more than likely, the company as a whole will benefit through enhanced staff understanding of the issues involved and the corresponding perception of all of the different aspects of the client relationship and its requirements.

9.3. SEEING INTO THE FUTURE

Such an analysis can then project alternate scenarios for change and development into the future and test whether the balance remains secure; if it does not and balance is lost, then change scenarios have to be adjusted so that the holistic balance within the company as a whole is preserved. This form of modelling can be used for benchmarking changes and for monitoring the company's holistic balance over time. Engineering consultancies, with their very high people and technology content, are well suited to systems analysis of this type.

These types of assessment put the people of a company right at the forefront of its strategic thinking and development; they reassure those who are employed in the company that their input not only counts, but is indeed essential, to the future of the company.

These kinds of tool can also be used, to good effect, to test acquisition and disposal strategies and to assess whether the 'fit' between companies is really genuine and meaningful, or whether it is simply just a 'good idea' on the part of management.

Companies are all dynamic entities; it is the responsibility of management to provide both the external and internal environment for the success of the company, and these types of assessments can only ever be an aid to that process.

NOTES FOR THE CHAPTER

CHAPTER 10

And where now?

Chapters 5 - 9 have looked at how market and practice requirements are likely to drive large firms of Consulting Engineers forward; many of these will apply to some extent to smaller and more specialist firms as well. As always, for the writer of a book like this, it is fairly easy to provide the analysis and provide pointers to some of the solutions; it's much more difficult for those who have the responsibility for putting these things into practice!

And, for every organisation, the eventual solution will be different, because it will have a different skill, people and experience mix to any other, as well as a different history and tradition. But, not only will it be different; no sooner has any new vision taken shape than it will need to evolve and change in response to the market, which is always dynamic.

10.1. THE PROCESS OF CHANGE

It is always difficult to change a company and the way in which it and its people work, because we all have a built-in affinity for what we know and the way in which we are used to doing things. Change usually has to be forced upon us. When this is the case, then it is better to change a lot quickly rather than attempt to change in very small steps; but there has to be some sense of continuity, some sense of order within the organisation, and these need to be identified, emphasised and built upon. The slow change process can so easily lose momentum and be defeated by the innate forces of conservatism.

And, in looking at areas of activity to change, there has to be a rigorous system that flags up non-essential activities - activities which, if stopped, *"won't mean the roof caving in"*.[1] The process will involve sacrificing many 'sacred cows' or activities which may have seemed, in the past, to be the back-bone of the company; it will require new attitudes to co-operation, both within the company and with peers and competitors. It will require restructuring the way things are done and the people employed and involved. Above all, it will have to put innovation, productivity and client satisfaction first; only then will profits flow that are robust and enable the company to continue to change and, hence, survive.

So as to direct the change process, *"the beacons of productivity and innovation must be our guide posts"* says Drucker in his book. Profits which do not reflect these, or which downgrade these, are not real profits; they are destroying the capital and intellectual worth of the company. Management, in making the decisions at this time, will often have difficult choices to make about the size of the business; sometimes, this can be quite an affront to its ego!

Staying still in today's world is not a realistic option, although it is sometimes tempting to stay still if at that particular moment the company's position feels comfortable. Businesses and their people have to carry on developing and changing; it is *'sink or swim - all of the time'*. One of the implications of this is that the senior management of engineering consultancies needs to be continually stepping back and doing reflective and strategic work rather than hands-on day-to-day monitoring, which has to be entrusted to the people of the company, with appropriate audit trails to ensure consistency and quality.

Estimates vary, but it has been suggested that up to half of a senior manager's time should be spent with peers and with people outside their own operating units; they use this time to collect personal information and new ideas. This is the source for new strategic thinking, which needs to keep going over the key questions:-

- What is the task?
- How are we going to accomplish it?
- Why do it at all?

---

[1] Managing for the Future. Peter Drucker. Butterworth-Heinemann, 1992

All of this represents very significant change for firms of Consulting Engineers, most of whom, until the late 1980s at any rate, had very traditional structures and ways of working. Changes have been forced on to the industry, but I suspect that there is still a long way to go in terms of completely altering attitudes to work and to each other, both at an individual and at a corporate level. Global competition and all of the new technologies will force the international firms to carry on changing quickly.

Managers of the future, with their new leadership skills, and with an openness to learn and work together, will always have in their sights the overall objective of freeing up their subordinates for creative work; they will act continuously as mentors for their ongoing development.

Overcoming fear of change is the main hurdle that many companies face when trying to change. Some guidelines are:-

- Dream/visions of the future. Try to use examples from other companies to stop people standing still. Make it exciting and evocative, but get people ready for it first, get attitudes right at the outset.
- Only start the change process when all are ready; laggards can stop it dead!
- Allow 5 years for a realistic programme. This seems a long time ahead, but you need a vision. This may not be completely right, but at least you start off in the right direction.
- Change by example. Be open, trusting, talking. Trust people to get on with it and let them 'own' it.
- Have milestones for the dream and regularly monitor against these.
- Build conviction that all can and will be achieved. Keep pushing forward - don't let the process pause or stop.
- Provide training readily and when asked for; this builds confidence.
- Beware of the 'conspiracy of silence'. Make internal politics dangerous. Get creative assent.

10.2. THE COMMITMENT OF ALL TO THE BUSINESS

Success in the present markets depends almost entirely on the skills and abilities of the people who work for an engineering consultancy, as well as their commitment and attitude to the business. They need to be empowered in whatever they do and feel that their particular contribution is both important and under their own control; then they will be able to also take responsibility for their own motivation and personal development.

Delivery on promises made by any single person in the company must be honoured by all; this is often a new concept! As I have said before, everyone needs to be convinced that changes are being made for the medium/long-term good of the consultancy; this ensures that they feel that there is an ongoing place for them in the business.

But the company must support this personal development, providing training, and the funding for training, where it's needed for the new working environment, for the training of our people is just as much an investment in the future as is introducing a new I.T. system. As the work-force becomes better trained and educated, so their expectations of involvement and reward will rise; they are also likely to want to then experience new challenges and learning experiences, thus will the cycle go on!

In many ways, these changes amongst staff members will mirror what has been happening for rather longer in the professions of accountancy and management consultancy. These professions are now well used to:-

• loose organisational structures
• project based internal and external relationships
• all forms of flexible working
• individual and team accountability, and
• reward systems that are just that, i.e. they reward performance, good or bad, rather than pay simply for attendance at work.

10.3. THE SMALLER CONSULTANCY

This book has concentrated primarily on the larger companies of Consulting Engineers, most of whom are international companies, because these are the firms which have sufficient weight and size to compete across a broad range of activities within the global market. With this size, etc. come all the problems of internal/external co-ordination, complexity of management and organisational structures, and the employment of a wide range of technical disciplines, for whom the market-place might dictate very different reward scales.

However, as indicated in Chapter 2, the smaller firms of Consulting Engineers should not be discounted from some of these impacts, although the surviving medium-sized   consultancy   is becoming an increasing rarity. For both of these categories of company, survival will require that they rigorously define their markets and concentrate on some of the changes which have been outlined in the book. They will always have the difficulty of obtaining and retaining key staff who are high-flyers; these people naturally gravitate to large firms for their wide range of activities and prestige, whereas the attraction of a smaller, perhaps more specialist, consultancy might depend much more on the remuneration package and personal development opportunities which might be on offer.

In contrast to the large firms, the smaller firms can develop a distinctive client portfolio which may not be in direct competition with the larger companies. One example might be a strong local presence in a small project market, trying to ensure that it is the personal connections that secure continuous turnover rather than the impressive range of skills that a larger firm might be able to muster from a distance. To achieve this objective, it will probably be necessary to have a very proactive alliance policy for the additional disciplines which it is perceived that others can offer; a small firm might even be able to act as the 'eyes and ears' for the specialist skills of a larger consultancy, but retaining that essential client contact and hence lead role and project control always closely to themselves.

An alternative approach would be for small firms to become so highly specialist that they develop a very widely spread client and geographical base. Then the boot might be on the other foot; it may be the larger companies that have the main client contacts and lead roles, but since they do not hold the particular specialism adequately within their own portfolio, they have to draw in the 'real' specialist from time to time. In this scenario, these types of specialisms might just relate to particular types of structures or technologies, for which there is truly a limited and diverse market that the large firms cannot always tap into effectively, for example, immersed-tube tunnel technology, specialist process or environmental modelling.

What seems to be fairly clear is that there is a business cycle whereby companies grow and become very much conglomerates within their industry; then they redefine their objectives and recognise that they can achieve greater value and performance commitment from their people by limiting the range of their activities; then they split into separate, often independent, operating units which simply co-operate with one another. Once management grasp this, and accept its inevitability, then their strategies for the development of a company can be geared accordingly.

It would not surprise me to find that some of the larger consultancies start to go this way, initially through a product-division split, followed perhaps by complete independence of part of the company as time goes by.

Thus, for the big firms, diversification will give way to specialisation, so as to retain their specialist staff and preserve their profile in the market and their reputation. Smaller specialist firms will learn to group together to offer all of the project skills that are needed, with alliances constantly shifting and, often, working together to find the best opportunities in the market.

It could well be possible that it is the smaller firms of Consulting Engineers, with their extensive use of the Internet and E-mail, who will become key underlying elements in supporting the larger companies. In order to do this effectively, they will have to ensure that their staff have many of the skills and attitudes that are just as relevant in the larger firms.

10.4. CLOSING COMMENT

The future for Consulting Engineers is clearly not going to be anything like what we have seen before. Perhaps firms of Consulting Engineers are going through the equivalent of a 'mid-life crisis' in their identities, realising where they have come from and perceiving that there are lots of new opportunities awaiting them, but that now is the time to grasp these, else they will be lost.

The changes are not going to be anything like we have seen before:-

- There will be the challenge of new forms of organisation.
- Creative thinking by all staff members will be the order of the day.
- Consultancies will need to continually engage their people's emotions and imagination so as to retain and motivate staff and thus preserve their intellectual capital, the real asset value of the business.
- All of their work will be knowledge-based, much will be I.T. and process-based. Forms of 'virtual working' are not going to be too far away into the future.
- Marketing will be strategically led, aiming at increased penetration and/or market domination in a restricted number of fields or clients.
- Workforces, working methods and relationships will be flexible, temporary and constantly changing.
- Reward systems and organisational structure will also be under constant change. Rewards will be for good work.
- Day/day management of the business, quality and control systems will become self-directed.
- Forms of organisation will need to be developed which will foster the continual growth of people's skills as these are relevant to future business. This will lead to continuous growth in expertise for the company concerned and help to create its platform of skills for the future.
- Many more staff will be involved in, and responsible for, the destiny of their company. They will also be responsible for client satisfaction.

- Physical location of staff, skills and client will become less of a constraint on development with the use of the Internet and teleworking.
- Work units may well become smaller, even completely independent financially of each other. Relationships between them will be very flexible.
- There will be a continuous search for relationships with others and for new work processes so that the extra 'added value' can be squeezed from a project; it is success in this area which will separate the 'winners' in the global race from those who simply survive.

Those companies that adopt a culture of continual and planned growth and change will win. Yet a tradition of continuity and common ownership will be an essential element in ensuring stability and internal confidence.

But sadly, I suspect, there are many who will not do these things - they may survive, but they will probably not prosper!

Yet however our future world develops, what is clear is that it will continue to need clean water, clean air, proper waste disposal, transport and power systems, new materials and new forms of buildings. These are all the province of the Consulting Engineer, who needs to be debating not only the 'how' questions but also the 'why' questions; for that is when people from outside the profession listen and take notice.

And as the 'why' questions are asked, and answered, then it will become apparent that the profession is not simply made up of technocrats; it consists of real people with assets, skills, views, and the whole range of human qualities. For, increasingly, it is that whole range which will be called upon to implement the work of the Consulting Engineer as we move dramatically into the 'information age' and unleash the talents of all.

National boundaries will disappear in this quest for excellence and survival. The successful firms will have the ability, by themselves or with others, to seek work anywhere and to deliver total projects to their clients, rapidly and economically.

"Engineers should stop talking to themselves and communicate with those who can restore their image. They must identify the issues, mobilise the players and start fixing the problems. But their voices will not be heard unless they step out of the crowd and up to the podium."[2]

---

[2] William Lewis, President of FIDIC, reported in Civil Engineering, Nov. 1997

NOTES FOR THE CHAPTER

APPENDICES

Company performance data

Appendix 1 - UK Consulting Engineers

Appendix 2 - European/Asian/N. American Consulting Engineers

Notes

1. These data are included in the book so as to give the reader some indication of comparative size and performance between different leading companies of Consulting Engineers. Accounting information is only ever a snapshot of the affairs of a company at a particular point in time and over, normally, a historical 12-month period; all such data need therefore to be read with some caution, and may not relate at all accurately to the present performance of the company.

2. In making international comparisons, a further difficulty is the use of different currencies and the fluctuations that occur in these. Generally, conversions have been made for currency values in October 1999.

3. Return on Capital Employed (ROCE) and Return on Equity are, so far as possible, quoted using the same parameters for each Appendix separately. Thus comparisons can be made within any one Appendix, but not between the two. Definitions of the precise calculation of these indices are, in any case, far from standardised; again, they are provided as illustrations which are, hopefully, helpful.

154

PERFORMANCE DATA FOR UK CONSULTING ENGINEERS

APPENDIX 1

Company	Accounts Date	Turnover £M p.a.	Staff Nos. approx.	T/o per head £000 p.a.	ROCE-%	Profitability % of T/o	Overseas % of T/o	Status/ Ownership
W.S.Atkins	03/99	430	8522	50	33	6.4	20	Plc
Ove Arup	03/98	217	3911	55.5	6	1.1	49	Ind
Mott Macdonald	12/98	191	3052	62.6	10	3	55	Ind
Hyder	03/98	139	2926	47.5		-1.2	50	Sub/UK
Halcrow	04/98	132	2545	51.9	11.5	3.4	65	Ind
Maunsell	06/98	131	3109	42.1	23.5	7.2	82	Ind
WSP	12/98	77	1501	51.2	30.7	6	10	Plc
Babtie	12/98	74	1734	42.5	28.2	5.2	small	Ind
Mont.Watson	12/97	70	1069	65.5		-0.7	56	Sub/US
Mouchel	07/98	54	1020	52.9	13.5	3.6	24	Ind
Gibb	12/98	44	934	47	12	3.6	65	Sub/US
Parkman	03/98	32	759	42.2	0.9	0.1	11	Ind
Oscar Faber	05/98	31	658	47.7	62.1	6.2	13	Plc
White,Young,Green	06/99	35	718	49.3	23.3	7.7	0	Plc
RPS	12/98	26	555	46.8	42.4	19	15	Plc
Waterman	06/99	30	609	49.8	20.4	8	3	Plc
Thorburn	02/99	22	448	49.4	27	3	9(1998)	Sub/US
High Point	07/99	25	366	69	40	9.4	68	Plc
MVA	12/98	21	358	58.3		-1	50	Sub/F
Aspen	03/98	16	323	50.3	24.8	1.5	3	Ind

Sources:- NCE Consultants File, Published Annual Reports, Proprietary Databases

PERFORMANCE DATA FOR EUROPEAN/ASIAN/N. AMERICAN CONSULTING ENGINEERS

APPENDIX 2

Company	Country	Accounts Date	Turnover M. p.a.	Staff Nos. approx.	T/0 per head £000 p.a.***	Profitability % of T/o	Return on Equity %	Overseas % of T/o	Ownership/ status
Arcadis	Netherlands	12/98	1275 NLG	7025	53	4.2	17.5	55	Public Company
Carl Bro A/S	Denmark	06/98	1200 DKK	2010	54	2.3	11.7	45	Foundation/Empl.
COWI	Denmark	04/99	1337 DKK	2099	57	-1.4		49	Foundation
DHV	Netherlands	12/98	507 NLG	3074	48	4.7	29	45	Foundation
Fichtner	Germany	12/98	143 DM	1100	42	1.6	17	60	Private/Employees
Haskoning	Netherlands	12/98	332 NLG	2042	48	3.1	14.4	43	Subsidiary
Holland Railconsult	Netherlands	12/98	275 NLG	1241	65	9.3		small	
LAWGIBB	U.S.A.	12/98	311 USD	N/A	N/A	4.7	21*	34	Private/Employees
Norconsult	Norway	12/98	600 NOK	820	61	4.4	14	24	Employees
Pacific Consultants	Japan	09/98	533 USD	1652**		N/A	N/A	small	Private Company
Ramboll	Denmark	12/98	1256 DKK	2026	56	2.3	16	20	Foundation
Stantec	Canada	12/98	185 CD	>2000	N/A	7.1	N/A	28	Public Company
Witteveen + Bos	Netherlands	12/98	110 NLG	608	53	10		small	Employees

Sources:-   Published Annual Reports supplied to Author

	Europe	Asia	N. America
Enquiries put out:-	12	3	9
Response:-	11	1	2

* Return on Capital Employed
** Full-time employees only
***See Note 2

NOTES TO REFERENCES

Chapter 4

Reference 3 - "Global work: bridging distance, culture and time" by Mary O'Hara-Devereux and Robert Johansen. Quotation supplied by Pritchett & Associates, Inc., Dallas, Texas, USA, in their publication "New Work Habits for a Radically Changing World" and used with their permission.

Reference 4 - Quotation from Warren Bennis supplied by Pritchett & Associates, Inc., Dallas, Texas,USA, in their publication "New Work Habits for a Radically Changing World" and used with their permission.

Reference 6 - "When Giants learn to Dance". First published by Simon & Schuster, London, 1989. Copyright Rosabeth Moss Kanter.

Chapter 9

Reference 1 - 'Springboard' - A Strategic Analysis Package. This is copyright of Stratology Consultants and described with their full permission; all rights are reserved by them.

Crown Copyright
References to HMSO publications. Crown copyright is reproduced with the permission of the Controller of Her Majesty's Stationery Office.

Butterworth-Heinemann
Butterworth Heinemann Publishers are a division of Reed Educational & Professional Publishing Ltd.

Pritchett & Associates
All quotations supplied by Pritchett & Associates are used with their full permission: all rights are reserved.

Pearson Education Limited
Both Prentice Hall and Longman Publishing are part of the publishers, Pearson Education Limited.

INDEX

160

Health & Safety, 73
Heller, Robert, 131
hierarchy, 59,111,117
High-Point Rendel, 41,154
Holistic Account, 42
        approach, 137ff,139ff
        Business Analysis, 141
        Operations, 42
Holland, 33
Holland Railconsult, 42,155
home computing, 55
hosting, 14,48
'hot desk', 56ff,132
human resources, 16,142
Hyder, 42,154

I

IBM, 54,55
incentive, 60ff
independence, 36
indigenous consultants, 11,19,25,70
industrial development, 11,68,84
Industrial Society, 61
information sources, 21,33ff
        technology, 11,40,62ff,84,85,
            101,122,125,126,130,148ff
infrastructure, 16,70,76,81
in-house teams, 3,14,65,82
innovation, 67,118
insecurity, 37
Intellectual Capital, 41,117,122,126ff,
            141,150
intelligence, 63
interim management, 48
internal competition, 51,112
        control, 113
        processes, 125

synergy, 113
international agencies, 11
        divisions, 114
        subsidiaries, 114
Internet, 55,130ff,139,150
ISO 9000, 72,73

J

Japan, 38,69,40,104
Johnson & Scholes, 87ff,111
joint ventures, 39,52
'just in time', 50

K

knowledge, 45,58,83,91,92
Knowledge Account, 41
KPMG, 112

L

landfill, 66
language, 68,69
Latham, 32,52,75,121
LAWGIBB, 39,42,154,155
lawyers, 52,75
lead consultant, 15,76
leadership, 50,58,59,80,98
lean organisations, 47
legal skills, 98
life expectancy, 108
local autonomy, 53
        availability, 18,70
Local Government, 14

risk assessment, 85
Road Construction Units, 9
Roffey Park, 50,94,105,107,110
role model, 7
role in society, 33
RPS, 42,154
Russia, 68

S

safety, 73ff,85
Scandinavia, 17,33,68
Scanlon, Brian, 135,136
secondment, 30
secrecy, 12,13,19, 28,38
self-confidence, 61
service, 133
  business, 5
'Shamrock', 53,119
shareholder value, 37
share ownership, 60,117,118
Shell, 110,
single-source procurement, 43
skills, 81ff,97ff
  matrix, 99
  rebalancing, 54
smaller consultancy, 148ff
social issues, 70
socio-economic, 81
soft technology, 81
software, 67,84,85
South-east Asia, 2,17
Southern Africa, 2,40
South America, 40
specialist/ism, 16,17,37,67,71,81,84,
          95,96,115,131,149
spread of offices, 25
'Springboard', 141

staff numbers, 23,27
  participation, 40
  retention, 39
  ownership, 58ff
stakeholders, 88,109,117,118,128,139
Stantec, 42,155
Statutory Accounts, 34
Stephenson, 7
stock, 24
strategic analysis, 88
  development, 140ff
  implementation, 90
  management, 87ff
  partnership, 82,97
  thinking, 51,106
strategy, 87ff
Stratology Consultants, 141
strengths and weaknesses, 90
subsidiary, 21,24,29,114
survival, 110
sustainable development, 11,66,81
SWOT analysis, 91,95
synthesising, 76,81

T

take-over, 3,101,106ff
targets, 128, 130
team approach, 39,59ff,62,80,107
team pay, 61,104,120
teamwork, 59,98,119ff
technical skills, 98,100
technology, 4,16,30,81ff
teleworking, 41,55,151
term appointments, 43
Thatcher, 10
Third Age, 48,53,54
Thorburn, 154

U

V

W